# OPTICS

*The Science of Vision*

# OPTICS
## *The Science of Vision*

### BY VASCO RONCHI

*Translated from the Italian
and revised by*
**EDWARD ROSEN**

DOVER PUBLICATIONS, INC.
*New York*

This Dover edition, first published in 1991, is an unabridged and unaltered republication of the edition first published by New York University Press, New York, in 1957. The work was originally published in Italian by Nicola Zanichelli, Bologna, in 1955 under the title *L'Ottica scienza della visione*.

Manufactured in the United States of America
Dover Publications, Inc.
31 East 2nd Street
Mineola, N.Y. 11501

*Library of Congress Cataloging-in-Publication Data*

Ronchi, Vasco, 1897–
   [Ottica, scienza della visione. English]
   Optics, the science of vision / by Vasco Ronchi ; translated from the Italian and revised by Edward Rosen.
        p.     cm.
   Translation of: Ottica, scienza della visione.
   Originally published: New York : New York University Press, 1957.
   Includes index.
   ISBN 0-486-66846-0 (pbk.)
   1. Optics, Physiological. 2. Vision. I. Rosen, Edward, 1906–   . II. Title.
QP475.R633   1991
612.8'4—dc20                                                    91-27433
                                                                     CIP

# *Translator's Preface*

Figure 22 is reproduced from Polyak's *The Retina* (University of Chicago Press) by kind permission of Mrs. Stephen Polyak. Figures 26 and 27 are reproduced from H. K. Hartline's article on "The Nerve Messages in the Fibers of the Visual Pathway," which appeared in the *Journal of the Optical Society of America*, 1940, *30*, 242, 244.

It is a pleasure to acknowledge my debt to two friends who volunteered to read the translation in typescript. Robert I. Wolff, Professor of Physics at the City College of New York, proposed a considerable number of improvements in the translation, which has benefited from his profound knowledge of physics, optics, and astronomy as well as from his love for precision in terminology. Lloyd Motz, Associate Professor of Astronomy at Columbia University, was likewise kind enough to offer some valuable suggestions. To both these men I desire to express my heartfelt thanks, at the same time absolving them of whatever imperfections remain.

E. R.

# *Bibliographical Note*

◇◇◇◇◇◇◇◇◇◇◇◇

Were I to recall the names and writings of all who have made some contribution to the studies utilized in this book, I should have to present an interminable list, so huge is the number of those who have dealt with this vast and important subject. I have therefore dispensed with the compilation of such a list.

V. R.

# Contents

◇◇◇◇◇◇◇◇◇◇◇

# OPTICS

*The Science of Vision*

# CHAPTER I

# The Definition of Optics

◇◇◇◇◇◇◇◇◇◇◇◇

**1.** As I sit down to write this book, about four decades have passed since I first heard about optics. Forty years ago I learned that there was light, consisting of rays capable of being reflected and refracted; that there were mirrors, prisms, and lenses able to produce images; that there were optical instruments; and that there was a sense organ called the eye. Certain interference fringes also were mentioned, and likewise a sequence of colors known as the *spectrum*. In the classrooms of the secondary school I thus formed an initial concept of optics: it was a chapter of physics.

A few years later at the university I had a second encounter with optics, again in the course on physics. There the ideas that had been outlined in secondary school were confirmed and perfected. The geometrical reasoning was refined and completed in that magnificent construction, the Gaussian homography. Light waves were studied in close connection with color, and their wonderful capacity to explain the phenomena of interference, diffraction, and polarization was demonstrated. The conception of the spectrum was broadened to include the ultraviolet and infrared regions. The electromagnetic nature of light waves was expounded. Knowledge of how the most important optical instruments worked was imparted. Everything was presented as beautiful, finished, orderly.

But then there was talk about a split. *Geometrical Optics*, which discussed rectilinear *rays*, was not in agreement with wave phenomena. It was, therefore, considered a provisional optics, useful and practical, but to be disregarded when an approach to the essence of the phenomena was desired. For then it was necessary to

3

enter into *Physical Optics*, which succeeded in accounting so well for the experimental data by means of the wave mechanism.

Inexperienced and timid young minds, impressed by the splendid results achieved by their predecessors and overawed by the authority of their professors, could not suspect the dangerous significance and profound importance of this split. In these minds the firm conviction grew, even without any such explicit assertion by the instructors, that by now in optics everything was known.

When I finished my program at the university, fate decreed that I should dedicate myself completely to the study of optics, in an environment of research in physics, where a special department devoted to optics for technical purposes was deemed essential. I then found out that there was also a *Technical Optics;* that there was an industry of *optical glass;* and that there was an *Engineering Optics*, busy with *optical calculations* for prescribing optical systems, and concerned likewise with designing the instruments in which these systems were to be placed.

I discovered that in addition there was a *Physiological Optics*, which undertook to examine the functioning of the eye regarded as an organ of the human body.

**2.** The further I advanced in the field, the vaguer my ideas became about the meaning of the term "optics." At first I had the feeling that optics, a very ancient and fully developed discipline, had as it were fallen apart, like a great empire that had dismembered itself by ceding a portion of its territory to each of its neighboring states, which had then proceeded to convert these areas into provinces of their own.

Since the possibility presented itself of creating an institution devoted exclusively to optics, I conceived the ambitious project of reuniting the empire. But as I deepened my knowledge of the subject, the problem seemed ever more indefinite. Optics was intermingled with so many different sciences that it appeared extremely hard to determine by a reasonable criterion where to put the border of the new empire. Indeed I ran the risk of incorporating in it units that were clearly alien to it or of leaving outside its frontiers districts that were unquestionably optical—the sort of difficulty that occurs quite frequently in drawing the political boundaries of nations.

This, however, was not a crucial question. What mattered was to work either theoretically or experimentally in such a way as to make real contributions to our knowledge; it was of little importance whether or not they came within the logical limits of the field bearing the name "optics."

**3.** The uncertainty about the meaning of this term, however, was a sign that something was wrong. Meanwhile in physics the famous controversy between the quantum theory and the wave theory became acute. Although its repercussions on the content of "optics" were only moderate, from the classificatory point of view it nevertheless led to the proof that *Physical Optics* could no longer be identified with *Wave Optics*. The latter became, after the indisputable establishment of the quantum concept, a new kind of *Geometrical Optics*. Just as classical *Geometrical Optics* had substituted a network of straight lines for the actual radiation, so now a system of wave motions was introduced. Thus when a treatment of the familiar phenomena of interference, diffraction, and polarization by the wave mechanism was desired, *Wave Optics* was invoked. But the task of seeking to ascertain the nature of the radiations had to be left to *Physical Optics*.

This clarification showed itself to be increasingly more accurate and sensible as research progressed. The meaning of the word "optics," however, suffered a new blow. For geometrical and wave optics, now deprived of a secure physical basis and reduced to the status of provisional studies of schematic models, like the ray and the wave, acquired the value of chapters of mathematics, from whose conclusions the necessary correspondence to experience could not be demanded, because it was admitted from the start that this correspondence did not exist. And if these two were "chapters of mathematics," they lost their standing as branches of optics, since the universal intention was to give this science the character of a chapter of physics.

Even more serious was the effect of the clarification on physical optics, so called. If optics was to be a chapter of physics, its essence should have been contained in physical optics. Instead, the latter lost all significance and ended up by disappearing altogether. For at heart its subject matter should have been the investigation of the nature and properties of the radiation. But this imposing task is a general theme of physics, because it concerns not only the visible but also the invisible radiations, whether these be electromagnetic waves, microwaves, infrared, ultraviolet, X rays or γ rays. The fact that a very small group of all these radiations is capable of affecting the eye forms a minor detail, utterly insufficient to justify the label of optics for so vast and important a class of studies. Its proper name is *radiation physics*.

**4.** The more I sought to delimit the meaning of the term "optics" and to define its content, the more they both escaped me. What was commonly considered most characteristic of optics as a chapter of physics slipped away into two purely mathematical constructions, to wit, geometrical optics and wave optics; and in the part that should have retained the essence of physical optics, general physics was found, and "optics" lost all its significance.

In the meantime, however, I had gone into the branches regarded as technical offshoots, which have as their aim to put the rules of optical science into practice.

One of the divisions of this technology is *optical calculation*. Anyone who examines it finds therein a variety of mathematical procedures, both algebraic and trigonometric, based on the laws of reflection and refraction. By means of these procedures the attempt is made to determine the curvatures, thicknesses, and diameters of lenses in order to find the proper combination. Anybody who concerns himself with optical calculation engages in an exclusively mathematical task, which is completed when he has compiled the "prescription" of the calculated optical system. This is, therefore, clearly a question of applied mathematics. It would be futile, however, to look there for anything distinctively optical. True, the point of departure is two postulates, namely, the laws of reflection and of refraction. But these two laws, it should be noted, are not peculiar to optics since all radiations and all wave motions, material included, obey them. The conclusion was inescapable that "optical calculation" was essentially a chapter of applied mathematics, wherein the only optical feature was the purpose for which the work was done.

The same inference had to be drawn when I went on to examine the technique of optical works, so called. I found first of all an industry that produced "optical glass" as a special and indispensable raw material for making lenses and prisms. But the manufacture of optical glass does not differ fundamentally from that of ordinary glass, except in the care that must be taken to ensure its homogeneity, and in the types or combinations of mixed powders, from whose fusion glass results. The technology of optical glass, therefore, is a glass technology, particularly refined, in view of the end for which its product is destined. But in itself it does not have any specific traits of its own, optically speaking.

This raw material, that is, the glass so produced, is *worked* by means of abrasives in order to give to the individual pieces the shape of plates, lenses, and prisms. This operation too does not have any specific traits of its own, optically speaking. It is a question of giving a geometrically determined form, generally plane or spherical, to the surfaces and then of polishing them. The same thing is done in many other processes and techniques, whether of wood, metal, stone, or glass. Therefore the working of glass to make lenses and prisms deserves the designation of optics only because the product is used for an optical aim.

Hence in the so-called optical technology, from calculation to raw material to finished product, I found no justification for the label other than in the intent. The calculating, glass-making, abrasion, and polishing are done as they are done elsewhere, with this difference, that when those calculations, glasses, and operations are to serve a purpose of another nature, they are not called optical; and they are called optical if the objects produced are to serve optical purposes.

**5.** I then proposed to define these "optical purposes." What seemed the most immediate and obvious purpose in common parlance was the production of "optical instruments." Again I asked myself what characteristics led to the classification of an instrument as optical, and I did not find even one. In short, I had to give up the idea that to consider an instrument as optical was "purpose"

enough to regard one of the aforementioned operations of calculation or fabrication as optical.

To sum it all up, my investigation had thus far led me to the
following observation. In ordinary speech there was talk about a
certain optics, which was supposed to be a chapter of physics; but
as such it disappeared in a very vast and much more general treatise, losing all its own specific traits. On the other hand, in that
presumed chapter of physics, studies developed that as usual were
distinctively mathematical. Or, to cross over into technology,
manipulations were carried out that pertained to mathematics or
mechanics or glass-making, but had nothing optical about them
except their purpose. In none of all this was anything found that
was intrinsically optical.

I had thus attained an advanced degree of perplexity, because
the conclusion was almost inescapable that a true optics *did not
exist*, at least in that mathematico-physico-technical sphere in which
I had looked for it with such great eagerness.

**6.** Then I bethought me that the world of physiologists, biolo-
gists, and pathologists also is interested in optics. In fact we speak
of *Physiological Optics* and *Ophthalmology*. Essentially, these disciplines study the functioning of the normal eye, the causes of ab-
normalities, and the diseases of the eye and their cure. It would
obviously be more appropriate to say "physiology of the eye"
rather than "physiological optics," just as we say, not "pathological
optics," but "pathology of the eye."

Whatever its name, this entire branch of study is in fact carried
forward in an atmosphere completely different from that described
above, with criteria and methods unrelated to the mathematical,
mechanical, and glass-making procedures, whose place in optics
was discussed in § 4. Moreover, the persons who pursue this new
class of studies differ in training and mentality from those in the
scientifico-technical group mentioned above. Between the two
groups there are sporadic contacts and tenuous relations. The students of physiological optics barely take into account that in what
they are accustomed to call "physical optics" (namely, the optics
considered by everybody to be a chapter of physics) there is "light"
consisting of rays or waves, and there are laws of reflection and

refraction that can be utilized to explain the functions of the cornea and crystalline lens in the mechanism of vision.

On the other hand, mathematico-physico-technical optics scarcely makes a fleeting reference to the fact that human eyes exist and people see the external world by means of these organs.

Accordingly, since optics was supposed to be a chapter of physics, I felt no urge driving me to seek the true essence of optics in the physiological sphere.

**7.** At the same time that these thoughts were evolving, my experimental activity moved steadily toward a new type of data, which almost all students of optics were instructed systematically to avoid. This activity started with the purely technical aim of taking theoretical rules, which were thought to be already established beyond dispute, and putting them into practice. The nature of the organization where this activity was carried on required me to furnish results decisive in settling affairs that were sometimes of considerable consequence.

In the beginning I met grave difficulties that were generated, not by special and incidental questions bearing on the apparatus for making the measurements, but instead by a kind of inadequacy in the laws of classical optics. Somehow I had a vague impression that the particular cases I had to deal with from time to time were almost never those to which the laws known to me, however general they might be, could be applied.

These laws did in fact concern the characteristics of the individual pieces of equipment or instruments subjected to my scrutiny. Those characteristics had to be defined unambiguously by means of numbers or exact units of measurement, which were supposed to be found with reasonable tolerances whenever the measurements were repeated in the presence of the interested parties, for one of whom it was advantageous that the numerical value should be higher, for the other, lower.

From the very outset it had to be recognized that in order to perform those measurements at least one observer was needed. Very frequently the results were the expression of his judgment, and therefore they were not as objective and unequivocal as they should have been to avoid undergoing dangerous oscillations when the measurements were repeated under test conditions by observers heavily interested in the outcome.

Between theory and practice there was this discrepancy. The theory defined certain magnitudes objectively and independently of the observer. In practice, on the contrary, in order to ascertain the numerical value of those magnitudes it was necessary to resort to at least one observer, and the result varied in accordance with his functional characteristics. This result did not yield the objective magnitudes that were sought, since it was at least in part subjective in nature.

This subjectivity had been noticed in numerous instances, and from many sides the proposal had been advanced to eliminate the living observer so far as possible, and replace him by various devices, like photoelectric cells and photographic plates.

Such substitution, however, was not always possible. Then the practice was instituted of defining a *normal observer*, capable of supplying reliable results. Unfortunately, however, the definition of normality was at all times rather vague, and it often depended on a statistical basis open to serious objection.

**8.** Under the impact of experiments conducted in the critical spirit of a man obliged to furnish definitive findings for which he had to answer with a deep sense of responsibility, I was led to view this procedure with less confidence than is currently placed in it. I was induced also to question the value of a definition of objective magnitudes, which could not be determined without the intervention of an observer, by reason of whose intervention the same result was not always obtained.

Thus I came increasingly to doubt the objective nature of these magnitudes, and to seek their origins or reconstruct the reasoning by which the founders of classical optics had proceeded to define them. Once I had entered upon this path, the dominating intervention of the observer's mind in all optical operations became clear to me, and I likewise saw plainly the herculean effort with which those founders had succeeded, not in eliminating the mind, but in deceiving themselves and others that they had eliminated it by a series of highly ingenious conventions.

Accordingly I was led (I say "led," for actually I resisted because of my intellectual disposition, and also as a result of my prolonged studies and of the training I had received) into the realm of the mind, and it appeared evident to me that it was utterly misleading to talk about optics without taking into proper account the intervention of this important, or rather absolutely predominant, factor. At first I was inclined to think that, apart from its technical aspect, optics should be regarded as being not only *physical optics* and *physiological optics* but *psychological optics* as well. Then I found that no mean body of work had been accomplished by specialists in this last field too, but it had remained practically unknown to the cultivators of the other fields of optics because of the considerable difference in their environment, point of view, and cultural background.

**9.** I was in this state of indecision and perplexity with regard to the definition of optics when, for entirely extraneous reasons, I was compelled to study ancient optics. I say "compelled" because, as a result of what I had heard said and seen written about the very slight value of that optics, I was so skeptical about the subject that of my own accord I would never have thought of undertaking to study it. Yet that study has clarified my ideas in an unforeseen and unforeseeable manner. The writing of this book is largely due to the influence exerted on my outlook by over two thousand years of work in the field of optics.

Contrary to the widespread opinion that optics began in the seventeenth century, I was surprised to find its roots in remotest antiquity. But without going back to so distant a period, about which adequate evidence is difficult to gather, I cannot refrain from mentioning two treatises of the fourth century B.C. that are still preserved in their entirety. Their author's name is Euclid (although it has been maintained that he is not the mathematician Euclid) and they are entitled *Optics* and *Catoptrics*. In them we already find the law of reflection, essentially equivalent to the law of today, even if expressed in a variant form.

The optics of those times was profoundly different from modern optics, if we apply the latter term to the optics that came to the fore at the beginning of the seventeenth century. Ancient optics was a science that lasted two thousand years. It deserves deep respect, even if it later fell and surrendered the field to a new optics. That has been and is the fate of all the sciences, including today's. And I hasten to add that not in all respects has the new optics marked an advance over ancient optics.

**10.** Ancient optics was a chapter of physics, even in those days. But the physics of that time was quite different from the physics of today. It was an *anthropomorphic physics*, a science in which the chief figure was sentient man. The chapters into which this physics was (and still is) subdivided reveal its origins beyond any shadow of doubt. Then too there was talk of mechanics, heat, optics, and sound; no great insight is required to see the connection with the *sensations* of force, warmth and cold, light and color, and noise. Ancient physics lacked the chapter on electricity and magnetism, for the very reason that the human body is not provided with the corresponding sense organs. It would perhaps be proper to say that ancient physics corresponds, in modern terminology, to our *physiology of the senses*.

In that early, yet magnificent, epoch of Greco-Roman philosophy, the key problem was the study of man as a sentient being. Thinkers sought to explain how humanity came to know the external world. They soon reached the conclusion that senses existed, which functioned through peripheral organs. These were linked by nerves to a central organ, the brain, where dwelt the soul or mind. The latter could come in contact with the outside world only by means of the sense organs and the signals reaching the mind by way of the nerves connected with the sense organs.

These signals, having been received and analyzed by the mind, are *represented* in specific ways. Thus those arriving along the nerves connected with the ears are represented by sound or noise; those coming in along the nerves linked with the various parts of the skin are represented as warmth or cold; those traveling along the nerves joined to the eyes are represented as light and color.

At that time there was a general conviction that sound, warmth and cold, light and color, like taste and smell, were *psychical representations*, entities created by the mind in order to represent the signals reaching it from the external world by way of the peripheral sense organs and the nerves connected with them.

The mind's capacity to create sounds, lights and colors, smells and tastes, warmth and cold was proved by the phenomenon of the dream. What is seen and felt in a dream, there is no doubt, is created by the mind that is dreaming.

**11.** The philosophers of those days went on to study the functioning of each sense. They undertook to find out how the individual senses learned about the properties of external bodies and transmitted this information to the mind. Thus, in the case of touch, it was quickly noticed that the communication between the outside object and the sense organ consisted of simple contact. In the case of taste also, it was soon discovered that for the flavor of a substance to be felt, it had to come in contact with the appropriate organs located in the mouth. In the case of smell, there is no direct contact. But for the odor of a body to be perceived, exhalations or vapors must detach themselves from the body and enter the nose. If this communication is prevented by some obstacle, the sensation of the smell is likewise interrupted.

In the case of sound, the communication is even more distant, but it too was identified. A vibrating body imparts its vibrations to the surrounding air, which transmits them to the ear. These are mechanical, silent vibrations. When signals arrive at the mind to the effect that the ear has been struck by these vibrations, only then is sound or noise perceived, since it is in this manner that the mind represents the group of shocks received by the ears.

Thus the problem of vision was reached, and here the road was not so smooth. The subject gave rise to a discussion in which the ablest philosophers of all the ages were to take part, but without decisive success. Only after twenty centuries was the solution attained that is still considered definitive today.

**12.** The separate peaks of this bimillennial effort will be explored in Chapter II, because they are not without importance for my thesis. Here I limit myself to a few fundamental comments.

In those two thousand years there were many studies of optics, which were mathematical, experimental, physiological, and even technical in character. The use of eyeglasses for the correction of vision goes back almost seven centuries, but that story does not concern us now. In their general lines these studies were carried on without a controlling and fruitful directive, that is, without a theory, for the simple reason that the mechanism of vision was not known; or rather, various notions were put forward which were so paradoxical and acrobatic that they did not aid the development of the studies. In fact, they had a decidedly negative influence on the acceptance of eyeglasses, and in the main turned out to be definitely sterile.

This swift survey indicates that the evolution of optical studies is closely connected with the theory of the mechanism of vision.

**13.** During the long period in which philosophers and mathematicians eagerly sought the key to this mechanism, one basic concept was clear to them all and beyond dispute. Light, what we see when we say "It isn't dark," is a purely subjective phenomenon, created by the mind to represent the external world. Since Latin was then used as the language of science, the term by which that light was plainly denoted was *lux*. The first fifteen of the twenty centuries in question talked only about this *lux*. If necessary, they added that *color* also had the same subjective nature. It too must be considered a creation of the mind, by which the latter represents some features of the outside world brought to its knowledge, again by way of the sense of sight.

The last five centuries, as we shall see in Chapter II, accepted the idea that *lux* was a subjective representation, but regarded it as somehow the effect of an external agent acting on the eye. The effort to identify that physical agent, designated by the term *lumen*, brought about one of the most important and most interesting developments in physics.

Even though the philosophers of the later Middle Ages had many profound differences of view with regard to other questions in the same field, none of them opposed this lucid and exhaustive position, which held the subjective *lux* to be the effect of the objective *lumen*.

It had also been made quite evident (and here it had even been exaggerated) that everything which was seen, precisely because it was a mass of figures created by the mind of each observer, was highly personal and subjective in character. To show that the thing seen was exactly as it had been seen and where it had been seen was not an adequate answer. For in rebuttal, this thoroughgoing skepticism brought forward a long list of "optical illusions," that had been minutely observed, analyzed and catalogued. They were the basis of that dreadful pronouncement, "Scientific knowledge cannot be acquired by sight alone," a pronouncement that has done so much harm to humanity.

Ancient optics, then, was definitely not objective and not physical, in the modern sense of the word, but had a purely subjective phenomenon as its foundation.

**14.** At the beginning of the seventeenth century the key to the mechanism of vision was found. But at the same time a revolution of vast importance took place, one of those catastrophic revolutions that overturn the deepest layers of human convictions. Man, who until then had been the lord of the universe, as it were, was dethroned. The earth, which was supposed to occupy the center of the world (or at any rate, of a world) and to have the heavens at its service together with the planets revolving about it, instead became a tiny part of a train of satellites around the sun, itself reduced to a modest component of one of the numerous nebulae.

Thus arose the concept of an immense universe, in which man formed an utterly negligible entity. For this little man nothing remained to be done in the field of science but to try to know the universe and discover its laws. The discipline entrusted with this task was physics.

It would surely be wrong not to recognize that, from the time physics was directed along this path, it has taken giant strides and transformed human life. But neither can it be gainsaid that this direction expresses, not a final and incontrovertible view of the world, but a philosophy just as subject to discussion as its predecessor. Above all, it should be clearly understood that the new physics was not a continuation of, and evolution from, what had gone before. On the contrary, it set itself in opposition to the previous outlook. It categorically denied having an anthropomorphic character, whereas the earlier science had been predominantly of that nature. Formerly the aim of physics had been to explain how the human mind came to know the outside world. Now the purpose of the new physics was to know the structure and laws of the external world, independently of the observer.

There would have been nothing peculiar in contriving a science of this sort, had it been newly established and not inserted in the ancient science, which it replaced in substance though not in form. People in fact continued, and still continue, to talk about mechanics, heat, optics, and sound, while the great chapter on electricity and magnetism is now added on equal terms with the other four chapters.

**15.** This organization of physics is strange because, as was indicated in § 10, the subdivision into those chapters had a meaning when human sensations were referred to, but lost it in a study "independent of the observer." Indeed the distinctions between those chapters steadily diminished until they disappeared—in substance, I repeat, because in form they are still flourishing.

Thus the physicists found that *heat* is a form of kinetic energy of the particles of the atomic world. They then decided to ignore the sensation of *warmth* and *cold*, and concern themselves only with heat, which thereby became a topic of mechanics. For this includes in its scope the study of motion, even if the moving bodies are of the order of magnitude of molecules or atoms. In fact mechanics not infrequently discusses the motion of material *points*, which are even smaller than atoms.

In exactly the same way, when it was discovered that sound was due to the mechanical vibrations of the solids and liquids by which they were transmitted to the ear, the new physicists resolved to study these vibrations and pay no attention to the sensation of sound. But in so doing they acted in a manner that can be explained only by recalling the revolutionary character of the philosophical movement of the seventeenth century. In all revolutions, political and scientific alike, there are always excesses. The victorious party maltreats the vanquished, destroying everything connected with it, including what was good.

Sound is without doubt a subjective phenomenon. Outside the mind there are vibrations. These, however, are not sound or noise, but a *silent* motion. Only when these vibrations have been received by an ear, transformed into nerve impulses, and carried to the brain and mind, only then, internally, is the sound created that corresponds to the external vibrations, and it is created to represent this stimulus as it reached the mind.

Had the physicists wished to act with crystalline clarity, they should have behaved as they did when, disregarding warmth and cold, they defined heat, which stands in the same relation to warmth and cold as a stimulus to a sensation. So they should have talked about acoustic vibrations, and avoided saying that these *are sound*. For the sound created by the mind is almost always assigned to a place outside. Hence to identify acoustic vibrations with sound may lead uncritical young people to believe that sound is actually a physical, and not a mental, phenomenon. It might be said that the physicists did not want to prevent this misunderstanding. For, as investigators of the world without an observer, they did not like to be forced to admit that their world was without sounds, and that if they wished to study sounds, they had to return to the mental world of the auditor. The successful attainment of their purpose cannot be denied, when we ask what concept of sound is acquired by students in schools all over the earth.

But if it is made clear that in the physical world there are no sounds but only mechanical vibrations (calling these vibrations "acoustic" is itself a reference to their capacity to stimulate an auditor's ear) sound ceases to exist as a separate chapter of physics, and instead is converted into a topic of mechanics concerned with vibratory motions.

**16.** I have tarried a bit over the case of sound because it offers us an excellent springboard for jumping ahead to the case of optics. Here too, when physics was transformed, a change of direction was effected, with the aim of studying chiefly the nature and properties of the external physical agent that had been called *lumen* in the later Middle Ages. But once more it was not comfortable to make it too obvious that the physical world was dark, or without *lux*, since that was present only in the observer's mind. Hence there was no further talk about *lux*. Latin had meanwhile been replaced as the language of science by the modern tongues. Each of them uses only a single word to denote an entity which, in the physicists' scheme, was supposed to be *lumen*, but which in the public mind ended up by being *lux*, because everybody almost always talks about what he sees. In English this word is *light*, in Italian *luce*, in French *lumière*, in Spanish *luz*, and in German *Licht*.

This linguistic detail is an interesting indication of the alteration in philosophical outlook that occurred in the seventeenth century, and it is also proof that the change did not constitute progress in all respects. For the overwhelming majority of students today are thoroughly confused about the meaning of the term "light." Most of the time they conceive it to denote that which is seen, the *lux* of our ancestors. But especially in scientific circles, when light is said to consist of rays or waves or photons, the intention evidently is to refer to *lumen*. Certainly almost everyone is convinced that the *lux* which he sees (I use the Latin expression because it is much clearer than the English) is physical, external, and identical with the *lumen*, while the observer plays no part whatever, in the sense that if he closes his eyes, the outside world continues to exist in all its brilliance and with all its colors.

**17.** For what has been remarked about *light* may be repeated about *color* too. The revolutionary physicists of the seventeenth century strove to deprive the observer's mind of any importance, and their undertaking would have been too difficult, had they stated clearly and explicitly that the world to which they wished to devote all their attention was without light and without color. So they kept quiet, permitting the illusion of a physical world full of light and color to dominate the neophyte's mentality. Nor can it be denied that they succeeded in their purpose.

Even today, when it is pointed out how far the confusion of ideas has gone, physicists usually reply: "It's a question of words." To what extent it is purely a question of words, the reader may judge for himself when he has reached the end of this book.

In the meantime, however, the new chapter on electricity and magnetism had undergone a magnificent development. From it electromagnetism arose, and that marvel was put together which consists of the great range of electromagnetic waves, those of enormous wave length like radio waves, of short wave length, like microwaves and infrared, and of very short wave length like ultraviolet, X rays and γ rays. In this last category were also the light waves, with all the colors of the spectrum.

**18.** The study of physical *light*, or *lumen* in the medieval sense, thereby entirely lost its specific character. If no consideration was to be given to the fact that this *lumen* was capable of affecting the specialized sense organs of an observer, then there was a relapse into electromagnetism pure and simple. Optics no longer had any reason to exist as a chapter of modern physics. Just as sound had become a part of mechanics, so optics was absorbed by electromagnetism. All of modern physics was thereby compressed into only two chapters, the *physics of matter* and the *physics of the aether*. The former consisted of ancient mechanics, sound and heat; the latter, of optics and electromagnetism.

In essence this arrangement was reached about a century ago. Had it been formulated lucidly and unambiguously, there need not have been any further talk in physics about optics, sound, and heat. Yet the misconception was, and is, so deeply rooted in thought and expression that a classification still continues to be used that has meaning only when the guiding principle of physical research— independence of the observer in general, and in particular of his mind—is rejected.

These are the main outlines of the conclusions to which I was led by the study of optics from Greco-Roman antiquity to modern times. The details of the subject will be taken up in Chapter II, where they will provide clear confirmation of the propositions just set forth.

**19.** The view that optics had lost its reason for existence as an autonomous science and should be regarded as a topic in electro-magnetism conflicted with the most elementary common sense. The cause of the conflict became evident as soon as the evolution of the ideas through the centuries was reconstructed. What had lost its reason for existence was *physical optics*, in the modern sense of that term. But anthropomorphic optics, or optics as understood by the ancients, a science which undertakes to discover the laws of that marvelous phenomenon, the perception of light, that is, of forms and colors, has ample reason for existence today, as in all ages past, and even more than in the past.

Optics defined in this way should be designated the *science of vision*. It is not a chapter of physics, nor of physiology, nor of psychology. It is a complex science that must take into account the contribution of all three of these disciplines. In every optical operation there is always a physical, a physiological, and a psychological phase. For a process to be truly optical, all three of these phases must be represented.

Therefore it is not correct to say that optics is a chapter of physics. On the contrary, it was this idea, based on some fundamental misunderstandings, as was mentioned in § 18 and as will be seen even better hereafter (§§ 51, 197), which led research down blind alleys and gave rise to many additional ambiguities and mistakes. Optics is not a chapter of a physics that aims to withdraw as much as possible from the observer. For optics has meaning when the aim is to discover the conditions and laws that permit an observer to see and see well. Optics as the science of vision is therefore anthropomorphic. It should not restrict itself to ascertaining the characteristics of the physical stimulus, but should concern itself with the effects of that stimulus on the sense organ, and with the consequences in the realm of the mind. For it is these last that are interesting, because they are really the end product.

This is the direction in which our analysis will proceed in the following chapters. We shall see that the significance of the innovation is much broader than may be supposed at first blush. To make the change in thought patterns and results more evident, I shall frequently institute a comparison with the accepted point of view and criticize it. The exposition will thereby acquire a slightly polemical tone, which I hope will not displease the reader. For the sole purpose of the criticism that I shall make is to attain a greater clarity of thought and a better knowledge of optical phenomena.

**20.** Before delving into the subject, however, I think it may be useful to delimit more closely the area that I propose to reserve for optics as the science of vision.

Undoubtedly standard optics also used the eye as an experimental means of research and check on the pronouncements of theory. Indeed the eye was widely used, not always with good judgment and the necessary care, as I shall have occasion to indicate repeatedly in the following pages. Yet in these activities the eye did not enjoy much respect, being considered an instrument that was a little too peculiar, unreliable, impressionable, in need of training, and not infrequently defective and *abnormal*. When it supplied responses in agreement with the theory that it was called upon to check, all was well. But if the responses were doubtful or negative, the fault lay with the eye that had furnished them, and not with the theory. Or else the conditions of the experiment had to be modified until the eye succeeded in seeing what it was supposed to see. The significance of these remarks will be made clearer hereafter (§§ 247–248).

More than a century ago experimental technique introduced other detectors or receivers, which according to the conventional mode of thought could with advantage replace the eye in the functions just mentioned. These other means of making comparisons and measurements were less "subjective" than the eye. Especially noteworthy among them were the photosensitive emulsions utilized in *photography*, and the photoelectric cells of various types for experiments with, and measurements of, *photoelectricity*. These new devices were (and in the opinion of many still are) incorporated in optics, further aggravating its already serious crisis.

It is really strange how easily the word "optics" is abused, to the point of talking about the "optics of X rays" and even the "optics of electromagnetic waves." It is strange because the people who employed these expressions (and they were persons of the highest rank in the hierarchy of science) never took the trouble to define what they understood by "optics."

**21.** Let us, therefore, try to view things in the most reasonable way available to us. We must first draw a picture of the physical world for ourselves. In accordance with the conclusions of the physics of matter and the physics of the aether, it is said nowadays that there are in the universe material bodies composed of molecules, atoms, and other primary particles, and that there are "radiations" or propagations of energy. Opinions about the structure of this radiation are not in complete agreement, because the construction of a mechanical model possessing all its known properties has not yet proved possible. In many discussions the wave model is used, and in others recourse must be had to the corpuscular model of *quanta* or *photons*.

Matter, then, emits and absorbs this *radiation*. There is only one way to detect it—absorb it and observe its effects on the energy of the matter absorbing it. The means of detecting it, so far as we are concerned, may be grouped as follows:

(*a*)  photoelectric cells
(*b*)  photosensitive emulsions
(*c*)  eyes.

The characteristic reactions by which these receivers indicate that they have absorbed radiation are the following:

(*a*)  for photoelectric cells, an electric current;
(*b*)  for photosensitive emulsions, a blackening;
(*c*)  for the eyes, perception of light and color.

**22.** Let us examine these processes.

In the case of a photoelectric cell, this is what happens. Matter emits radiation, which reaches the cell and is absorbed by it. An electric current is thereby generated, and in a circuit connected with the cell it deflects the pointer of a measuring instrument, e.g., a galvanometer. What is "optical" about this whole process? Obviously nothing.

In the case of a photosensitive emulsion, this is what happens. Matter emits radiation, which reaches the emulsion, is absorbed by it, and brings about changes in its atomic structure, thereby modifying its chemical properties. This layer of the emulsion is then treated in a reducing bath, as a result of which grains of metallic silver are deposited in the areas where the radiation had acted. This deposit of silver produces the *blackening* of the emulsion by increasing its capacity to absorb any radiation that may be directed at it thereafter. What is "optical" about this whole process? Obviously nothing, until someone looks at the photograph. But that is a later operation and concerns the finished photograph, not the taking of the photograph.

**23.** We thus arrive at the very simple conclusion that photoelectricity and photography are not optics. Indeed we may go further. We may consider the radiation, not by itself, but in intimate connection with the receiver-detector. The process of detection then becomes the subject matter of three sister sciences, which should be called:

(a) photoelectricity

(b) photography

(c) optics.

I say that these "should" be the labels, and quite probably they will be in the course of time. But at present there is a deep-rooted habit of including photography in optics. To be practical, then, and to avoid being misunderstood by many people, we may conveniently employ a provisional nomenclature, using the name "photographic optics" for what could be termed simply "photography," and "visual optics" for true optics, the science of vision, which utilizes the human eye as the means of detecting the radiation.

This trichotomy has a profound philosophical significance. In the first place, as was pointed out in § 21, the presence of radiation is indicated to us only by the process of detection, which consists of absorbing the radiation and transforming it into other manifestations of energy. It is, therefore, proper that the receiver-detector, which has so important a function, should exercise an influence also on the classification and terminology.

In the second place, any discussion of the radiation, its structure, and its properties, without reference to the means of detection, clearly takes on a completely hypothetical and conventional character. If it is proposed as a philosophical commandment to talk only about what can be demonstrated experimentally, the detector and the effects of the radiation on it must come first. This is a very thorny topic, to which we had better return near the end of the book (§§ 270–271).

# CHAPTER II

# The Basis of Seventeenth-Century Optics

◇◇◇◇◇◇◇◇◇◇◇

**24.** In § 12 I referred cursorily to a long, intense scientific effort which, in the course of two thousand years, succeeded in establishing the fundamental concepts of "light" and the "optical image" as well as in discovering the key to the mechanism of vision. At that time I did not go into detail because I promised to return to the subject and give it full attention. It is in fact a topic of basic importance for understanding the significance and implications of ancient optics and of the seventeenth-century revolution that led to the construction of the then new optics. It is very important also because it will permit us better to comprehend the foundations of this seventeenth-century optics.

In order to explain how the organ of vision acquired knowledge about the properties of the bodies in front of it, ideas of the following kind were put forward in antiquity: every alteration that is produced or received takes place as the result of a contact; all our perceptions are tactile, all our senses being a form of touch; hence, since the soul does not go forth from within us to touch external objects, these must come to touch the soul by passing through the senses; but we do not see the objects approach us when we perceive them; therefore they must send the soul something that represents them, a likeness, *eidola*, shadows of a sort or images that cover the bodies, move about on their surfaces, and can detach themselves for the purpose of transporting to the soul the forms, colors, and all the other qualities of the bodies from which they emanate.

The philosophers of those days followed a logical line in their thinking. Because they refused to admit the possibility of *action at a distance*, obviously some communication between object and sense organ was necessary. Hence they said that all our senses are a form of touch.

**25.** For the senses of touch and taste, the communication was a direct contact; for smell, it was an exhalation, a vapor; for hearing, it was a vibration transmitted by the surrounding air. For sight, there could be no appeal to a contact, because bodies are seen even when they are very far away from the eyes. Nor could there be an appeal to an exhalation or a motion conveyed by the circumjacent medium, because these means of communication allow only one effect to be felt at a time. One sound is heard at a time, one odor is smelled at a time. If two exhalations blend, the odor smelled is the resulting combination, just as if two vibrations mingle in the air, a single sound is heard, the product of the two vibrations conjoined.

On the other hand, in the case of sight, the forms and colors of innumerable bodies are seen *at the same instant* with wonderful sharpness and precision. Hence if there is something that moves from the observed body toward the observer's eye, it cannot be a shapeless exhalation, but must be something carrying the form and colors of the body whence it emanated. Therefore it must be like the body. Since it cannot be the actual body, it must be a skin that leaves the body, bearing what the body's surface presents to our vision, namely, form and colors. This is the reasoning that led to the conception of the *eidola* or skins or images or, as they were called later on in the Middle Ages, *species*.

Up to this point there would be no serious objection. The continual emission of these images in all directions might seem not impossible, by analogy with what obviously and demonstrably happened in the case of odors. But the first grave problem that presented itself was to explain how the images of a body as big as a mountain could enter the pupil of the eye, barely 2 mm in diameter.

To solve this problem, another quite acrobatic property had to be attributed to the images. They were supposed to contract along the way until they became small enough to enter the pupil of the human eye. This property seems even more peculiar if we bear in mind that the receiving pupil may be located at any point whatever, near or far. The images that started out in any one direction thus had to diminish at different rates in order to be able to fit into a pupil wherever encountered.

**26.** Even so, new difficulties arose at every single step. For example, despite all the acrobatics of this hypothesis, the question of the distance at which an object was seen still remained unsolved. The images, assumed to be appropriately reduced, entered the eye, impinged upon the "sensorium" or sensitive surface, and delivered to it the data pertaining to the form and colors. The data were transmitted by way of the optic nerves to the mind, which thereupon proceeded to construct a representation by means of a figure possessing that form and those colors. But there is an infinite number of such figures, all of different dimensions; and besides, the figure had to be located in the surrounding space at a definite distance. Obviously the images did not carry along with them any information suitable for determining these two geometrical factors, the distance and the dimensions.

If the followers of this theory went a little further into detail, things grew more and more intricate. It was known, for instance, that when a person looked at a plane mirror, he saw behind it the figures of the bodies in front of it. It was not hard to imagine that the images rebounded from the mirror's surface, deviating in exact accordance with the laws of mechanical reflection. But one complication constituted an unforeseen and insurmountable obstacle. The figures seen behind the mirror are not congruent with those that would be seen by looking directly at the corresponding objects, but are symmetrical to them with respect to the reflecting plane. Why do the images in rebounding from a plane mirror become transformed into their symmetrical counterparts? This was an enigma.

In addition it had to be explained why nothing could be seen in the dark; why an object, even if very small, looked blurred when it was too close to the eye; why certain minute things were not seen, like a needle on the ground, then suddenly were seen; and a thousand other such perplexities.

Those who supported this theory, in reality few in number, disposed of most of the difficulties by setting them aside without even mentioning them.

**27.** If the supporters of this theory were few, there were on the contrary very many who adhered to the opposing system, championed by highly able and well-known mathematicians. This latter doctrine evidently took its cue from perspective, in which the eye constitutes the "point of view." On the physical side the theory started from the fact that a blind man, even without touching an object directly with his hands, could become acquainted with its form by exploring it with a stick. Similarly, it could be supposed that from the eye there went forth rectilinear sticks, capable of examining the external world and bringing to the mind data suitable for knowing and representing the world's forms and colors. These straight lines that came out of the eyes were called *visual rays* and could be reflected as well as refracted.

This construction outdistanced the other a little in degree of absurdity, and the difficulties it met in explaining many circumstances of vision were no less. Why were the visual rays unable to study the outside world when it was dark? Why could they not probe a body very close to the eyes? When they were reflected by a plane mirror, why did they cause a figure to be seen behind the mirror instead of in its proper place? How did they go about reaching distant bodies like the sun, moon, and stars?

This list could be extended quite a bit. These defects were pointed out, not by me, but by adversaries at the time. The discussion of the subject was often lively. For everybody, to demolish was easy; but when it came to constructing, that was another story.

**28.** Aristotle too came to grips with these questions. For the two theories mentioned above were not the only ones. Some compromises were also attempted that sought to extract from each theory its reasonable elements and to discard its most acrobatic features. But nobody achieved any noteworthy result. Aristotle took an entirely different position. Concisely criticizing the theories of images and visual rays, he demonstrated the absurdity of both, and launched another idea that came close to the mechanism of hearing.

He brought back into the discussion not only the medium inter-vening between the object and the eye, but also the object's action on this medium culminating in perception by the eye. His language is very obscure, however, and he sticks to generalities so much that he made no real contribution to the solution of the problem, and attracted few followers in this field.

On the other hand, the theory of visual rays had rather a long life. To witness its decline, we must come all the way down to the eleventh century of the Christian era. Before we go into this new period, which was marked by the pre-eminence of the Arabs in general and Ibn al-Haitham in particular, we should observe that up to that time "light" was not spoken of in the modern sense of the word. In the physical world only images or visual rays existed, depending on which theory was accepted, while *lux* occurred in the realm of the mind. Nobody thought of a physical entity comparable to the electromagnetic waves or photons of today. In the philosophi-cal vocabulary "it is light" indicated a state of affairs, just as the expression "it is dark" did, and still does nowadays.

**29.** Ibn al-Haitham, who was known to the West as Alhazen, made a fundamental contribution that was a real stroke of genius. He dealt a mortal blow to the theory of visual rays with the phe-nomenon now called an *afterimage*. A person who looks at the sun and then closes his eyes, Ibn al-Haitham pointed out, continues to see the solar disk for some time. Furthermore, while gazing at the sun, the observer feels pain. These two facts definitely conflict with the mechanism of visual rays. For if the emission of these rays entailed suffering, they would not be emitted. Also, as soon as the eyes shut, vision should stop.

The actual phenomena, on the other hand, require an external agent that impinges upon the eye. When this agent is too strong, it afflicts the sensitive organ, and leaves impressions on it that may last quite a while.

The theory of visual rays did not survive this shrewd thrust and was considered outmoded. The theory of images leaped into the forefront. But as it was then constituted, it too could not withstand the attacks to which its flank was exposed. Ibn al-Haitham deserves undisputed credit for so modifying it that it assumed the form which was later to develop into seventeenth-century optics.

His first change consisted of providing a mechanism that took away from the images their property of contracting along the way before reaching the observer's eye. If the object was smaller than the pupil, Ibn al-Haitham reasoned, its tiny image could be propagated in a straight line and enter the pupil, wherever encountered, without any need to be reduced in transit. Any object, whatever its size, was to be thought of as resolved into many minute units or elements or points. Each such point emits in all directions its own images, which can enter a pupil, wherever they meet it, without having to undergo any alteration along the way.

**30.** In another astounding burst of genius, Ibn al-Haitham went on to explain how all these point images, having entered a pupil, could reconstruct inside the eye an orderly assemblage resembling the object that had emitted them. Referring to the structure of the eyeball as it had been described by Galen, he observed that all the coatings were to be considered spherical and concentric. When a point image strikes the outer coating or cornea, its incidence may be either perpendicular or not. If it is perpendicular, the image continues straight on into the interior of the eye. If the incidence is not perpendicular, the image is deviated by refraction. According to Ibn al-Haitham, an image deflected by refraction must lose its power to stimulate the "sensorium." Hence this is affected only by those images that reach the cornea perpendicularly; in other words, by only one for every point of the object in front of the eye. These special images thus enter the interior of the eyeball, intersect at its center, and then redistribute themselves over an extended area, preserving exactly the order and arrangement they had when they left the object.

By this device the insertion of a big, complex image in the diminutive pupil of an eye lost all that unnaturalness which had made the theory of *eidola* so objectionable in previous centuries. The new doctrine is not yet perfect, but the step taken toward the ultimate solution is truly gigantic.

**31.** Ibn al-Haitham himself felt that the situation was not entirely satisfactory. He went looking for difficulties even where they did not exist. He noticed, for example, that the order in which the point images were arranged after intersecting at the center of the eyeball was, to be sure, identical with the order in which they had been emitted by the object, but inverted. For the figure corresponding to the object came out upside down at the back of the eye. In those days it was taken for granted that the function of the eye was simply the receipt of information from outside to be transformed into physiological data and transmitted to the mind, which had the task of reconstructing the external representation. Nevertheless to a physicist like Ibn al-Haitham, the inversion of the figure at the back of the eye seemed absurd because it should have led to inverted vision of the corresponding object.

He tried to avoid this inversion of the figures inside the eyeball and hit upon an idea that nobody liked. He saw that if the sensorium were located on a surface in front of the eyeball's center, there would be no inversion. And since in the eye diagram that he used, this condition was satisfied only by the anterior surface of the crystalline lens, he was induced to conclude that this, although completely transparent, had to be the surface sensitive to the impressions coming from outside.

Besides these weaknesses, there are not a few contradictions to be found in Ibn al-Haitham's work. They are in all probability due to the evolution of his thought, a process inevitable in anybody who worked as long as he did on a subject still in an undeveloped state. Even so, his influence on the problem of vision has been decisive, both as regards the formation of the figures inside the eye by a mechanism involving point images, and also as regards the existence of an external agent able to act on the eye.

**32.** The fact that a person feels pain when looking at the sun made Ibn al-Haitham think that the *solar rays* must consist of something capable of affecting the sensorium to the point of hurting it. He further ascribes to the rays the power of making point images leave bodies when these are illuminated by the sun. With a number of extremely interesting arguments he succeeds in convincing himself that this physical agent or *lumen* must exist, and he also undertakes to ascertain its nature. The idea that the rays of *lumen* are the trajectories of minute material corpuscles is already expressed in his work. This is the first time that there is a discussion of the entity which physicists today call "light," and the first time that the corpuscular theory is presented. It is highly significant that Ibn al-Haitham describes macroscopic mechanical experiments, that is, experiments performed with material bodies of ordinary dimensions, in order to show that such bodies are reflected and refracted in a manner closely resembling that of the rays of *lumen*.

Of course Ibn al-Haitham's book is not confined to the few comments made about it above. I called attention there only to its salient features, which affected the development of optical theory. Before going any further, I should indicate that he too devoted no inconsiderable space to the description of *optical illusions*. He thereby confirmed the opinion of his predecessors that the eye as a sense organ was often subject to error and heavily influenced by the observer's preconceptions and mental faculties. For this reason the eye had to be regarded with a certain distrust, because to see a thing was not absolute proof that it really existed, or that it existed as it was seen.

This was a note that dominated ancient optics. The verdict cited above in § 13 that "Scientific knowledge cannot be acquired by sight alone" for over twenty centuries was looked upon as an indisputable verity. It had philosophical and practical reverberations of incalculable extent. "Optical illusions," so numerous and manifest, were always at hand to demonstrate the truth of that horrible judgment.

**33.** Let us return to Ibn al-Haitham. In his time culture was at a low ebb in the West. But when an interest in science revived there, the Arabic masterpieces were translated into Latin. In this way Ibn al-Haitham's treatise became known to thirteenth-century writers on optics, including Roger Bacon and John Peckham of England as well as Witelo, who was of mixed East German and Polish descent. Through their publications, particularly Witelo's, the work of Ibn al-Haitham was diffused in the West.

The results were completely negative. The theory of visual rays lost all credibility, but mathematicians continued to talk about it, partly because it lent itself so well to discussions of perspective, and partly because it was expounded in the schools. At the same time, however, the theory of images also experienced great difficulty in becoming established, for Ibn al-Haitham's analysis in terms of points was too difficult to understand.

What ensued was an indescribable atrophy of thought. Ideas tended more or less to cluster about the doctrine of *species*, a new edition of the ancient *eidola*. These *species*, however, were produced by the *lumen*, when it impinged upon a body, and they moved along the observer's visual rays as though along rails guiding them toward the eyes. During this motion they contracted in order to be able to enter the pupil. The contraction no longer constituted a serious obstacle because Ibn al-Haitham's mechanism had provided a sort of justification for it.

In other words, an effort was made to combine the classical with the new. The merger was a monstrosity, with which the philosophers and mathematicians of the later Middle Ages tried to reason when confronted by optical problems.

**34.** The outcome is revealed to us in a highly interesting and significant way by the attitude of those scientists toward eyeglasses. These small disks, which like the common lentil were thicker in the middle, began to be used in the thirteenth century to correct presbyopia. With their polished and transparent surfaces, they possessed the mysterious power of making old people see well at close range after having lost that ability with advancing age. They may have been invented accidentally by a glassworker who was fashioning disks for another purpose. Perhaps he was an elderly artisan who, while testing disks by looking through them, noticed that he saw the figures of nearby objects as clearly as in his youth.

When the invention was made known to the philosophers, it was examined by the standards of the prevailing theories and decisively rejected. Could any other verdict have been reached? Eyeglasses were transparent, to be sure, but they caused refraction and deformation. Whether the analysis operated with visual rays or *species*, obviously eyeglasses could be regarded only as a disturbing factor. For it was neither logical nor sensible to bend the rays that came out of the eye for the purpose of exploring the external world, and thereby make their task harder and perhaps falsify the result. Nor was it logical or sensible to deform and thereby disturb the *species* while they were hastening toward the eye with the intention of bringing it the forms and colors of outside bodies.

The decision was couched in a more practical and experimental form, but it was just as final and unappealable: the aim of vision is to know the truth; eyeglasses make figures look bigger or smaller than they would be seen with the naked eye, nearer or farther away, at times distorted, inverted, or colored; hence they do not make the truth known; they deceive and are not to be used for serious purposes.

Today this judgment elicits a smile, but at that time effective arguments against it could not be discovered. In fact the entire philosophical and scientific world disregarded eyeglasses. That they were not abandoned and forgotten is due to the ignorance and initiative of modest craftsmen, whose minds were not preoccupied with the reasons for things and with theories. These men found eyeglasses to be a useful device, constructed them, applied them, and made a living from them. To detect any change in the situation, we must come all the way down to the seventeenth century.

**35.** In the meantime the study of optics developed, but in the absence of the key to the mechanism of vision the advances were rather slight. There were two directions in which noteworthy gains were made, the mathematical and the experimental.

Although operating with visual rays, the mathematicians of antiquity had succeeded in formulating the law of reflection in geometrical terms, applying it to mirrors both plane and curved, and deriving interesting conclusions from it. The *Catoptrics* ascribed to Euclid is a little book devoted to just this topic. But, be it noted, it does not refer to "images" in the sense in which the term is used today. The origin of that concept will be dealt with presently (§ 51). It had no place in the mechanism of visual rays.

The investigations along these lines were therefore pure "mathematical games" since they had no connection with experimental reality. Indeed, if they had any effect, it was but to delay the system achieved later. For instance, when the reflection of rays by a spherical surface was studied, an entire hemisphere was considered, not a segment of a sphere. That is quite natural. Even today the formulas for the area and volume of a whole sphere (or hemisphere) are known and remembered by many, while few recall those pertaining to a spherical segment. Moreover, the behavior of rays on a hemisphere lent itself to very simple and conclusive demonstrations, whereas it would have been very difficult with the mathematical equipment of those times to determine with the precision characteristic of that science what was supposed to happen on a segment.

**36.** When a bundle of rays parallel to the principal axis (§ 145) of a hemispherical concave mirror strikes the mirror, the reflected

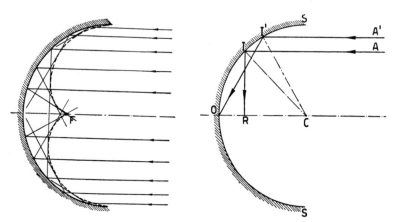

FIG. 1. Caustic formed by reflection     FIG. 2. Reflection in a concave mirror: two special cases

rays envelop a *caustic* (Fig. 1), which is a surface of revolution with a cusp *F* on the axis at the midpoint of the radius. Today this caustic constitutes an *aberration*. Then it did not, for it was the genuine and foremost phenomenon. It is interesting to notice by what direct and irrefutable demonstrations the argument advanced, demonstrations that we take good care not to repeat nowadays.

Thus if in Fig. 2 the ray *AI* is parallel to the principal axis *CO* and incident upon a point *I* such that the straight line *IC* joining *I*, the point of incidence, to *C*, the center of curvature of the mirror *SS*, forms an angle *ICO* of 45° with the axis, then *AI* must be reflected so as to be perpendicular to the axis. For the angle *AIC*, being equal to the alternate interior angle *ICO*, is 45°; hence by the law of reflection the angle *CIR* also is 45°; therefore *IRC* must be a right angle.

Again, if the ray *A'I'* is likewise parallel to the axis and incident on a point *I'* such that *I'C* forms an angle *I'CO* of 60° with the axis, the reflected ray must be *I'O* and must pass through the vertex *O* of the mirror.

Other instances of this kind, which could be treated by the geometrical methods of the time, were analyzed too. But the general case leading to the definitive conclusion may as well be given at once. Fig. 3 shows any ray *AI* parallel to the principal axis of the hemisphere. *IR* is the reflected ray, and *IC* is once more the straight line joining *I*, the point of incidence, to *C*, the center of curvature. It is evident that the triangle *IRC* is always isosceles. For the angle *RIC* or *r* is equal to the angle *AIC* or *i* by the law of reflection; the angle *ICR* or *i'* is equal to *i* because they are alternate interior angles; hence *i'* = *r*;

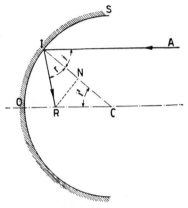

Fig. 3. Reflection in a concave mirror: the general case

therefore the perpendicular *RN* drawn from *R* to *IC* always cuts *IC* midway between *I* and *C*.

We may now reverse the reasoning. If *A* is a ray parallel to the axis and incident upon a point *I*; if *I* is joined to *C*; if at *N*, the midpoint of *IC*, a perpendicular is erected meeting the axis at a point *R*; then the straight line joining *I* to *R* is the reflected ray. By this very easy construction the reflected ray could be obtained for any incident ray. Besides being thus enabled in effect to trace the section of a caustic, these mathematicians reached the conclusion that all the rays reflected by the hemisphere and corresponding to an equal number of incident rays in a quadrant like *OS* (Fig. 3) cut the axis at various points between the midpoint of *CO* (the point *F* in Fig. 1) and infinity behind the mirror.

It is obvious why these elementary and elegant geometrical demonstrations are no longer repeated in books on optics today.

But it is also obvious that those mathematicians, who clearly understood how a hemisphere reflected a bundle of rays like those coming from the sun or a star, could not discover any connection between that phenomenon and the figure of the star which they saw when they looked in a concave mirror turned toward the sky.

Judged by the style of reasoning in optics today, those mathematicians went completely astray.

**37.** The other direction in which studies of optical phenomena proceeded was the investigation of refraction with the aim of arriv-

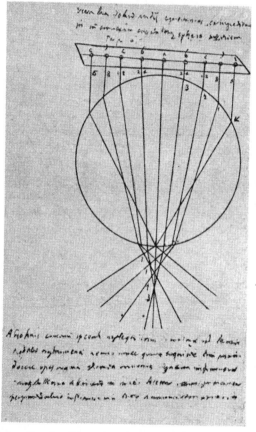

Fig. 4. Refraction through a glass sphere, according to G. B. Della Porta's unpublished manuscript *De telescopio*

ing at the formulation of a law, an effort that continued for centuries without accomplishing anything. The earliest description of a refraction experiment occurs in the *Catoptrics* attributed to Euclid. Claudius Ptolemy proposed a complicated equation, which served no useful purpose. Many experiments continued to be performed, but did not succeed in finding the clue to the intricate puzzle.

In general these experiments consisted of passing a beam of sunlight through an opening so that it fell obliquely on the surface of water contained in a glass vessel, thus permitting observation of what happened at the bottom. Here something very complicated was seen. Besides a spot nearly elliptical in form (the beam emerging from the aperture being sensibly conical) reddish and bluish colors appeared. At that time ideas about color were somewhat vague.

Then refraction at a curved surface began to be studied, with attention naturally turned first to the whole sphere. Sensibly parallel rays, like those obtained by filtering a beam of sunlight through an opaque screen perforated by some holes, were projected on a sphere (Fig. 4). These experiments followed the lines of the mathematical results previously obtained for hemispherical mirrors. Once more the conclusion was that parallel rays which entered a glass sphere, on emerging from it, intersected its axis at many different points. The hemisphere also was analyzed, and the outcome was the same.

Again, as in the case of spherical mirrors, no link could be found between the path of the rays studied in this way and the figures seen by looking through the sphere.

**38.** The scientists who experimented with whole spheres and hemispheres of course took good care not to do so with eyeglasses. It would have been foolish to complicate things with segments when simple spheres yielded no results. Accordingly, eyeglasses went on living exclusively in artisan circles. There the additional invention was made of correcting myopia with lenses that had concave, instead of convex, surfaces. When this was done is not known, nor by whom.

We have seen in essence what the state of optical knowledge was at the end of the sixteenth century. The field was marked by the greatest confusion, despite the earnest philosophical and experimental effort of two thousand years. The fundamental cause was the lack of a theory of vision, without which optics could not organize itself and go forward.

**39.** To show how difficult it was under those conditions to describe even the most elementary experiments (for us today this is a highly useful exhibit) I shall quote a passage from Giovanni Battista Della Porta's little work *On Refraction*. In this book, which is of very considerable historical importance, he put the following heading over a proposition: "An object seen perpendicularly within a medium denser than air enters unbent; but if seen obliquely, it deviates from the perpendicular." He is referring in particular to a body immersed in water. According to his statement, things should go like this: if the eye looks at an object in the water along a line perpendicular to the surface, the object leaps out without deviating and enters the eye; if this, on the other hand, looks along an oblique line, the object leaps out but deviates from the perpendicular in order once more to enter the eye.

This is undoubtedly a very strange way to depict the process of refraction at the surface of water. Obviously Della Porta did not mean that the object literally came out of the water, but he did not know what else to say instead.

**40.** The clumsiness of his terminology, which unmistakably betrayed the confusion and defectiveness of his concepts, may be further illustrated by another proposition. This is worth citing for the purpose of accustoming the reader to be critical of expressions used even today without adequate attention to the meaning of the words.

The proposition reads as follows: "When the refracted image of an object meets the eye, it is not seen in its place." Now the situation is no longer what it was in the proposition quoted above in § 39. What moves toward the eye this time is the image (not in the modern sense, but rather a likeness or *species*). Therefore the object remains in place. Yet matters are still far from clear. This image moves toward the eye and of course enters it. Where is it seen? Inside the eye? According to Della Porta's explanation later, he means that it cannot be seen where the object is. This agrees with everyday experience. But, as is evident, Della Porta does not know how to say all these things, simple as they may be. The reader may well imagine, without needing to inspect further samples, what sort of verbiage spills out when Della Porta goes on to talk about slightly more complicated optical phenomena. But the study of these herculean efforts on the part of an observer who tries to look without theoretical preconceptions at what he sees makes the power of those preconceptions manifest.

Yet in the sixteenth century a level was attained marking the commencement of a new and extraordinarily productive period, when that seventeenth-century optics was created to which I have already referred so often.

**41.** Its beginnings are found in two short discussions by Francesco Maurolico of Messina. The first was finished before he was thirty, the second not until he was sixty; a dozen years later he raised additional questions in an appendix. Yet despite their immense value, these writings were not printed until a generation after his death in 1575. His books in several other fields were readily published in his lifetime and widely sold. Perhaps he was too far ahead of his contemporaries in optics. How much that science's progress was retarded through the limited circulation of Maurolico's thought in manuscript instead of printed form is difficult to estimate.

To him we owe (at least so far as I am aware) the idea that rays emanate in all directions from every point on a body. This is a basic concept, by no means obvious and self-evident. It perfects, or so to say, purifies the masterly and invaluable teaching of Ibn al-Haitham (§ 29), according to which every body must be regarded as consisting of an infinite number of point elements, each emitting little images of itself in all directions. The trajectories of these images are the rays. Therefore innumerable rays leave every point on a body in all directions.

Prior to Maurolico, Western mathematicians did not reason in this way. Partly on account of a natural tendency (even now it is very hard to persuade pupils at school to apply the point-by-point analysis to objects instead of viewing them as a whole) and partly on account of the influence of ancient Greco-Roman philosophy, Westerners were attracted to the complex and complete *species*. It may not be without significance that Maurolico, born at Messina in 1494, was the son of an Oriental physician whose family had fled from Constantinople to escape the Turkish invasion.

Thus there now begins to be talk about geometrical rays leaving the points of self-luminous or illuminated bodies. These are rays of *lumen*. Maurolico's brief book contains some other interesting

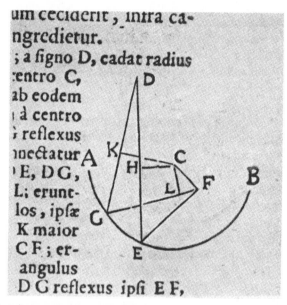

FIG. 5. Maurolico's diagram illustrating the convergence of rays reflected by a concave mirror

features, which are barely outlined. He draws a diagram, for example, to illustrate a concave mirror reflecting two rays which converge at a point (Fig. 5). This is a startling innovation, when compared with the beliefs of his time about the behavior of spherical mirrors.

**42.** Maurolico was, however, an isolated pioneer who was not understood. His activity was like a preliminary breaking of ground to prepare for an enduring edifice. The builder of this structure, which raised all of optics to a higher level, was Johannes Kepler. In 1604 he published a *Supplement to Witelo (Ad Vitellionem para-*

*lipomena*), but it was a supplement that made his medieval predecessor a forgotten man. The ideas contained in the *Supplement* provide the basis for seventeenth-century optics. By reason of their importance to us, they will be scrutinized with special care.

Taking his cue from the writings of Della Porta, as he explicitly admits, Kepler proposes to find the key to the mechanism of vision.

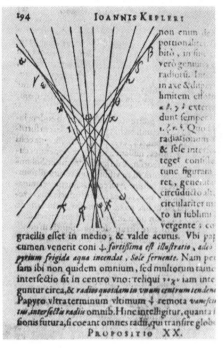

FIG. 6. Kepler's diagram illustrating the concentration of refracted rays near the cusp of the caustic

His thinking proceeds as follows. External bodies consist of aggregates of points. Each point emits in all directions rectilinear rays that are infinitely extended, unless they meet an obstacle. Considered by itself, a point is like a radiant star. If there is an eye in front of this point-star, all the rays that enter the eye will form a cone having the star as vertex and the pupil as base. These rays, refracted by the cornea and internal parts of the eye, go on to make a new cone whose base is again the pupil but whose vertex is a point on the retina.

Kepler arrives at this fundamental conclusion by studying refraction in a *sphere of water* according to an equation of the type

$$\frac{i}{r} = k$$

where $i$ is the angle of incidence, $r$ is the angle of refraction, and $k$ is a constant. He regards the formula as valid for angles not greater than 30°, and for the value of $k$ he takes $4/3$ in the case of water.

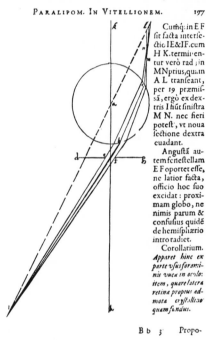

FIG. 7. Kepler's diaphragm

In describing the behavior of the sphere of water, he introduces another extremely important innovation. When a cone of rays is incident upon the sphere, the emergent rays cut the axis at many different points, to be sure. But if we observe the emerging group, in the cusp of the caustic we see a concentration of rays passing virtually through a point, whereas the others are rather far away (Fig. 6). Therefore, instead of allowing the whole sphere to function, we may limit it to a part by means of a very narrow diaphragm (Fig. 7). In this way, if a cone of rays is incident upon a segment of the sphere, the corresponding group of emergent rays will form a cone with a point vertex.

**43.** This was the first time that the function of the pupil was interpreted correctly. Before then almost everybody had stumbled over the problem of making the *species* enter such a small aperture, and had thought that everything would be so much simpler and easier if only the pupil were bigger. Ibn al-Haitham also was aware of the difficulty due to the dimensions of the pupil. In his case, too many images emitted by the same point object entered the pupil, and so he went searching for a sieve in the mechanism of refraction (§ 30). Thereafter nobody took up the question again until Kepler. He discovered a new and very important service of the pupil: it is a diaphragm. It delimits a portion of the eyeball and cuts off the disturbing peripheral rays that do not unite to form the refracted cone.

Kepler thus established that the rays emitted in a cone by a point object, after entering the pupil, reconverge at a point on the retina. That is where the stimulation of the sensorium occurs, and where the signals originate that are transmitted to the brain and mind. It then becomes the mind's duty to represent the received signals by creating a figure having the form of a point or luminous star, and to locate it where the object is.

**44.** This last question is examined by Kepler with truly extraordinary acumen. In order to locate the luminous star, the eye must be able to determine the position of the point object in space. The direction of the rays arriving at the cornea from the point object is linked to the position of the retinal point which receives the stimulation. For if that direction changes, the stimulated point changes too. Hence the mind has a way of identifying the direction in which the point object must lie. But the object's distance from the eye has to be ascertained also.

The solution of this difficult problem is found by Kepler in the cone of rays that has the point object as its vertex and the pupil as its base. He coins the term "distance-measuring triangle" for the triangle that has its vertex in the object point $S$ and its base in a diameter of the pupil (Fig. 8). In other words, he assumes that the eye is able to perceive the divergence of the rays forming the two long sides of this triangle. His conclusion is that the mind locates the luminous point at the vertex of the cone of rays reaching the cornea, or, what amounts to the same thing, that the luminous point is seen at that vertex.

For an extended object, the foregoing reasoning is repeated point by point, so that on the retina there is a figure resembling the object in all respects. The observer's mind, informed by the signals reaching it along the optic nerve, reconstructs the external figure point by point, and locates it at the distance and in the direction indicated by the distance-measuring or telemetric triangles of the

FIG. 8. Kepler's telemetric triangle

individual points. In short, the figure seen is the external projection of the figure intercepted on the retina.

In these rules may be recognized the foundations of modern geometrical optics.

**45.** Kepler applies them at once in an astounding manner. He explains how the figures of objects in front of a plane mirror are seen. For the first time this familiar yet mysterious phenomenon is exhaustively elucidated. His analysis has been reproduced in the opening pages of all the books on optics in the past three and a half centuries.

The rays emitted by a point object $S$ are reflected by the mirror $MM'$ so as to form a cone with its vertex at a point $I$, symmetrical to $S$ with respect to the mirror (Fig. 9). When the reflected rays reach the eye, the mind utilizes the telemetric triangle and locates the luminous point or star in the symmetrical point. For an extended object this reasoning is repeated point by point, leading to the conclusion that the figure reconstructed by the mind and located behind the mirror is not congruent with the object, but symmetrical to it with respect to the reflecting plane.

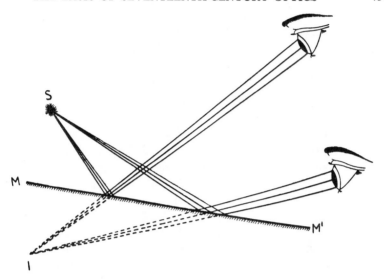

Fig. 9. Image of a point in a plane mirror, as explained by Kepler

After so many centuries of bafflement and confusion Kepler's explanation seems simply marvelous.

**46.** When he experiments with transparent spheres, Kepler finds that all goes well, i.e., in close conformity with his theory, if the pencils of rays are intercepted on a screen. But if they are received directly by the eye, what is seen does not submit to a simple, uniform generalization. He attaches so much importance to this difference of behavior that he suggests two different names. The figures intercepted on a screen are to be called "pictures," while those seen directly by the eye when it looks at mirrors, prisms, and lenses will be termed "images of things." This is a distinction of great value, which has never been appreciated as much as it deserved. We shall soon have occasion to return to it (§ 51).

We have now reviewed the main features of Kepler's first masterpiece. Seven years later another momentous contribution was to follow. For in his *Supplement to Witelo* Kepler does not deal with lenses. He mentions them briefly in one passage where he explains how they correct presbyopia and myopia. In this connection he points out that the function of eyeglasses is to vary the con-

vergence or divergence of pencils of rays so as to put directly on the
retina the vertices of the cones refracted by the cornea. But that is
all, for the present.

**47.** In that same year, 1604, a Dutch maker of eyeglasses began
to produce telescopes with a diverging eyepiece. These were copies
of an Italian model of 1590 that had been brought to Holland.
Among the learned the invention met the same distrust as had been
customary with regard to lenses in general for over three centuries.

The telescope thus remained in the hands of glassworkers, an
uninteresting device, feeble in effect, condemned in philosophy,
uncomprehended in practice. But this time the period of noncom-
prehension lasted only five years. In 1609 Galileo heard about the
instrument, made one for himself, and understood its enormous
importance for science. That was the dawn of a new scientific era.

Imbued with a *new faith* that what he saw in the telescope was
*true*, even if nearer or farther away, bigger or smaller, than what
was seen with the naked eye, Galileo made his celebrated astronomi-
cal observations. Of these the most sensational and revolutionary
was the discovery of Jupiter's satellites. He promptly published his
findings in March, 1610, in a remarkable little book entitled the
*Sidereal Message.* The entire academic world reacted violently, with
one voice accusing Galileo of extolling as real discoveries figures
seen only with the telescope, a notoriously misleading and un-
trustworthy contraption.

A controversy of colossal proportions broke out. On one side
stood Galileo, alone but unshakable in his faith and conviction; on
the other side, all of conventional science, permeated by skepticism
and a spirit of negative criticism.

**48.** Questioned from all sides, Kepler kept quiet, for he too was
perplexed. Finally in August, 1610, he laid hands on a telescope
made by Galileo, who had sent it to the Elector of Cologne. Kepler
carried out observations with the mental disposition of a man intent
on destroying, but he ended up agreeing that Galileo was right.

Kepler was thus the second scientist filled with Galileo's faith in
the telescope. The rest of the scientific world was still hostile and
distrustful. But victory was now assured.

Lenses were about to make a dramatic entrance on the stage of science, and would very soon become a topic of current and general interest. Kepler was the first to concern himself with them. In a matter of weeks he worked out the theory and printed it in his *Dioptrics* of January, 1611. He could do so because he had at his disposal the foundations of the new optics, which he had published seven years before. After more than three centuries of empirical life among the artisans and under the ban of science, lenses acquired a mathematical theory and became scientific instruments.

This was one of the most momentous and catastrophic revolutions recorded in the history of science. It is really amazing that so stupendous an event is practically unknown. For it meant the establishment of a new faith, which radically altered the attitude of the scientist and research worker toward observational instruments. Formerly the skeptic was unwilling to look through them from fear of being deluded by appearances. Now the insatiable investigator pushes a device's potentialities to the limit, seeking to obtain from it information, even fragmentary and deceptive information, about the macrocosm and microcosm.

This change of attitude opened a boundless horizon to scientific research and progress.

**49.** The revolution of 1610 shattered classical optics, causing it to vanish entirely from view. Today it is so completely unknown that anyone who proposes to read a work on optics earlier than 1600 must first make a special study like a student preparing to peruse a book on an unfamiliar science. Yet the transition from the classical to the new optics was not immediate. So profound a change not only in the customary canons of reasoning but even more in intellectual outlook could not occur from one day to the next. On the contrary, it was necessary to wait for the mature to pass away and be eliminated by the inexorable law of nature, and for young minds to be shaped in the new direction.

In fact, some years after the appearance of the *Dioptrics* those who understood its contents were very few. Half a century later, when its ideas and hypotheses no longer had any rivals (the *species* having disappeared from scientific discussions of optical subjects) they were still regarded with considerable reserve. But as one generation followed another, Kepler's concepts succeeded in acquiring unchallenged standing, and in the course of time a standing even higher than they really deserved.

It became a fundamental and undisputed rule that every point on a body emitted an infinite number of rays in all directions, and

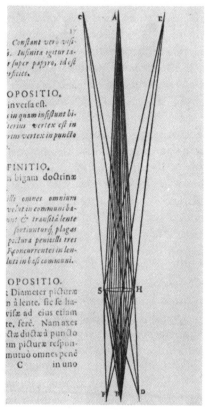

Fig. 10. Divergent and convergent cones of rays in Kepler's *Dioptrics*

that the divergent cones of rays which impinged upon a lens were transformed into convergent cones with their vertices so arranged that taken together they reconstituted a figure similar to the object (Fig. 10). Even now nobody would dare to question this rule, had it not been made clear that the model of "light" consisting of rays was inadequate. Apart from this reservation, which may be eliminated by hypothesis as in geometrical optics, Kepler's reasoning is regarded as an unalterable foundation.

**50.** The success of Kepler's doctrine was abnormally great because it found an extremely favorable reception in the dominant philosophical movement of the seventeenth century. At that time, as was remarked in § 14, a current of empiricism set in which ran ever stronger and stronger to the point of exaggeration, and as a result there was also a fierce reaction against the general tendency of medieval thought. It was an age when anyone with a suitable argument at his disposal shouted from the housetops a boast of having proved that "the Masters had erred." Everybody applauded him and said he was right. Sometimes he was wrong.

In the field of our interest the impression prevailed that after centuries of atrophied theory, the truth had been discovered. As a matter of fact things went wonderfully well in optics as compared with what had gone before. In any system of thought those "species flying through the air" were a rather objectionable feature. Now, on the contrary, the basis of the reasoning was a simple geometrical mechanism in marvelous agreement with experience.

Equally marvelous were lenses, which more and more every day extended the possibilities of research in the heavens and in practical applications here below. Then there was also microscopy, which further contributed to increased admiration for the triumphs of optics. And, be it noted, the microscopy of the seventeenth and eighteenth centuries was carried out mainly with simple microscopes, that is, single lenses. The followers of the new faith in direct vision and in vision with lenses could rightly reproach their medieval predecessors (and also the ancients, who already had the concave mirror, another form of the simple microscope) for having by their skepticism and ignorance deprived humanity of so many centuries of applied microscopy.

**51.** In this frame of mind scientists could not fail to extend an enthusiastic welcome to so useful a model of *lumen*, amenable as it was to the application of geometrical propositions that had been fully developed for centuries. In fact, the entire attention of students of optical matters was concentrated on that model.

On the other hand, Kepler's rule of the telemetric triangle, according to which the eye *had* to see every luminous point at the vertex of the cone of rays reaching the cornea, sanctioned a lack of interest in the mechanism of vision. This consisted essentially, so far as its physical aspect was concerned, of the eye's capacity to measure the small angle in the telemetric triangle, and to determine the direction of that triangle's axis. *The adoption of Kepler's telemetric triangle, therefore, permitted the physiologico-psychological aspects of vision to be ignored.*

Nothing could have been more agreeable to the philosophers and scientists of the seventeenth and, above all, of the eighteenth century. The incorporation of Kepler's rule in the empiricism of the time was responsible for the disappearance of two fundamental distinctions, that between *lumen* and *lux*, and that between "images of things" and "pictures."

*Lumen* and *lux* were merged in "light" (§ 16). "Images of things" and "pictures" were combined in the concept of the optical "image." There now arose the purely formal distinction between *real image* and *virtual image*. The intimate relationship, on the one hand, between real images and "pictures" and, on the other hand, between virtual images and Kepler's "images of things" is obvious. The former pair are both intercepted on a screen. The latter pair are both *seen* (or rather, are said to be seen) at the vertices of cones of rays reaching the eye, when those vertices are on backward prolongations of the rays reaching the eye. Apart from the fact that there are also "images of things" which today are called real images (we shall discuss that subject fully in Chapter IV below) it should be made quite clear that these two things were sharply distinguished by Kepler with his wonderful insight. He counseled ignoring the figures seen when the rays are admitted directly into the eyes, and urged concentrating attention on the "pictures" instead. To have deemed this distinction useless and to have eliminated it was considered "progress" by the successors of the pioneer who found the key to the mechanism of vision.

**52.** We must recognize that the course of events was also best suited to bring most effective support to this behavior by the students of optics. For once thought was centered on the *lumen* of old, the subject proved to be so fascinating that it attracted to its orbit the most eminent and most famous scientists of the seventeenth century.

When the problem of determining the physical nature of *lumen* was tackled, Ibn al-Haitham's hypothesis of a corpuscular structure encountered serious difficulties, since it conflicted with common sense in not a few typical phenomena of light. A stream of students, probably composed of many, not always brilliant, schoolmasters and as such convinced Peripatetics who have remained anonymous, revived those vague expressions with which Aristotle had barely adumbrated a theory of vision and, being guided most of all by the analogy with sound, little by little worked out a wave structure of *lumen*. Thus arose that profound wave-corpuscle dualism which is not yet resolved.

Two events concurred to re-enforce the position of the new enthusiasts for optics regarded as a part of physics. Reasoning with a *lumen* consisting of projectiles, Descartes in 1637 published the precise law of refraction (which had, however, been grasped previously by Snel[1]). Thus another puzzle two thousand years old was solved in the new theoretical climate.

**53.** The other event concerned color. Kepler had avoided talking about it. According to the opinion prevailing in the first half of the seventeenth century, light was white and colorless. Sunlight was pure and perfect. When it impinged upon bodies, it besmirched itself, so to say, and came away indued with color.

But when this operation was subjected to experimental control, a mass of contradictions and inconsistencies ensued. Light falling on a sheet of red paper came away red. But when it was incident upon red glass, why did it become red if it passed beyond the glass but remain white if reflected back? And why did white sunlight, transmitted through a prism of the clearest glass or purest water, emerge spread out in strips of various colors?

---

[1] This spelling with a single "l" is not a misprint. For the correct form of the Dutch surname, and the reason for the common mistake concerning it, see p. xiii of *The Appreciation of Ancient and Medieval Science During the Renaissance* (University of Pennsylvania Press, 1955) by George Sarton, whose recent death on March 22, 1956 was an irreparable loss to the history of science.

Innumerable mysteries of this sort were encountered in the endless experiments performed on the subject. One of the most in-
scrutable of all these enigmas was the coloration of thin plates. Why
did a fine layer of a white substance like soapy water become tinged
with so many different colors when illuminated by pure sunlight?
Soap bubbles readily showed this phenomenon. A related fact
was observed when two flat plates of highly polished glass were
brought into contact with each other.

After Descartes took the first step toward the solution of this
intricate question, a decisive clarification was contributed by
Francesco Maria Grimaldi. By a most beautiful demonstration he
reached the conclusion that "color was not located on bodies, as
was generally believed, but was a modification of the structure of
*lumen.*" He also suggested conceiving color as due to a vibration
transverse to the trajectory of the rays of light.

This was another stride on the road to the establishment of
optics as a part of physics. Color, on which the old-fashioned philosophers still held a tenacious mortgage by defying anybody to
deny its absolutely subjective character, now became a "modification of *lumen*" and therefore passed over into the physical world.

**54.** Grimaldi reported the discovery of a new property of
light, which, however, he did not succeed in explaining. It
was a new property, but also a new mystery, which he called
*diffraction.* He observed that when *lumen* passes through small
holes or slits, and when it grazes narrow obstructions like needles
and hairs, it gives rise to a number of strange phenomena, mani-
fested as lighter or darker and also colored fringes, where least
expected.

At the same time another puzzle came along to intensify the
interest of physicists in the nature of *lumen.* The law of refraction
had been enunciated; its validity and universality were further
confirmed every day by new measurements. But in accounting for
it physically, a multitude of obstacles was encountered that no one
would have anticipated. One fine day a naturalist named Rasmus
Bartholin discovered a crystal called "Iceland spar," which exhibits a special refraction. To an incident ray, two refracted rays
correspond, one of which obeys the ordinary law of refraction

strictly, while the other pursues an entirely different path and under certain conditions even disappears.

The very simple and classical phenomenon of reflection likewise presented its own puzzles. Transparent surfaces, like those of glass and water, also reflect a small percentage of the incident *lumen*. Why? Nobody succeeded in finding a plausible explanation of this curious effect. If *lumen* is allowed to fall on a glass plate having plane surfaces, reflection is observed to occur equally at the surface of incidence and the surface of emergence. In the former case, it occurs at the glass and within the air; in the latter case, at the air and within the glass. In the former case it is possible to discover a reason in the fact that, glass being denser than air, a part of the *lumen* finds it difficult to penetrate within and therefore returns to the air. But in the latter case the difference in density is the reverse of the situation in the former case.

Then in the latter case air was replaced by water, which is much denser than air. The result was exactly the opposite of what was foreseen. The beam reflected by the water into the glass is almost imperceptible, being much weaker than when in place of water there was air. This was another mystery.

**55.** By conceiving a *lumen* of rectilinear rays Kepler had opened the door to a series of puzzles. The "natural philosophers," who had undertaken to investigate the laws of the physical universe, were called upon to answer embarrassing questions that grew more numerous from day to day.

It is no wonder that in the face of such a situation the problem of vision, now regarded as solved, no longer aroused any interest. When Isaac Newton, by applying the conception of gravitation to the particles constituting *lumen*, not only explained reflection and refraction but also discovered dispersion, he reduced rather than increased the number of mysteries. Hence it is no wonder that from all (or almost all) sides a sigh of relief was breathed, and Newton's work was hailed as one of the masterpieces of science.

Although he succeeded in establishing a correspondence between the colors of the spectrum and the masses of the corpuscles constituting *lumen*, Newton was quite explicit and precise about the subjective nature of color:

The homogeneal light and rays which appear red, or rather make
Objects appear so, I call rubrifick or red-making; those which make
Objects appear yellow, green, blue and violet, I call yellow-making,
green-making, blue-making, violet-making, and so of the rest.

He clearly indicates that if at any time he speaks of light and rays as
colored, he must be understood to be speaking "not philosophically
and properly, but grossly." "For the rays to speak properly are not
coloured. In them there is nothing else than a certain power and
disposition to stir up a sensation of this or that colour," just as it is
the trembling motion of a bell or musical string that makes us hear
sound.

Although Newton was so outspoken (and he made these statements in 1704, when he published his *Opticks*) his followers disregarded his formulation. For them color lost all its subjective character and became a purely physical quality of "light." They forgot
that when their master called rays "red," he meant "red-making";
and they convinced the entire world that he had proved the rays
were red.

**56.** Light, color, and images thereby all became physical entities. To the observer no other function remained than to act as
a receiver and apply the rule of the telemetric triangle. It is possible
that the observer may sometimes err in applying the rule or, in
general, in exercising his receptive faculties. That is no concern of
the physicists. For if the observer errs because he is distracted, so
much the worse for him. Let him be more attentive and keep a
sharp lookout. If he errs because his organism is defective at some
physiological or psychological stage, that too does not interest the
physicists, for they must deal only with normal people. Let the
others seek the help of a physician or psychiatrist.

No wonder that after this change excellent experimenters like
Bouguer set themselves the task of "measuring light." Thus
*photometry* was born. No wonder that a century later the project of
"measuring color" was proposed, giving rise to *colorimetry*.

**57.** The march of events down this slope was quickened by the
spread of specialization in science. Since physical, physiological, and
psychological factors enter into the visual process, it is very complex.

At a time when the sum total of human knowledge was much smaller, men could deal advantageously with its entire range. To do so required an extraordinary intellect, but more than one was found. As experimental research evolved, however, the total grew so vast that it is no longer humanly possible for a single individual to grasp it all even if he spends his whole life at the task. Thus specialization was born, a development deprecated by many but nevertheless unavoidable and becoming ever more widespread.

In the field of scientific optics (if we leave technical optics to one side) there are clearly three divisions, to wit, physics, physiology, and psychology. Physicists are trained, as is well known, in a mathematical or technical environment; physiologists, in a medical environment; and psychologists, in a humanistic environment. The three groups differ in their intellectual outlook, cultural background, and very terminology. Optics was, therefore, bound to be dismembered, and the mutilation was carried out in drastic fashion. The physicists made of optics a chapter of physics. The physiologists dedicated themselves to the study of the eye as a sense organ, without taking into account the nature of the luminous stimulus, and without entering the mysterious and shadowy realm of the mind. The psychologists devoted but little time to the study of visual phenomena, although some experimental investigations were carried out.

On the physical front, problems of great importance and general interest were tackled by first-rate scientists who achieved wonders. But on the physiological front, after the magnificent victory over the mechanism of vision, enthusiasm slackened and the results of research lost their universal appeal as they were directed toward the pathological rather than toward the physiological sector proper. On the psychological front, the contribution was modest, as has already been mentioned.

Inevitably, the physical approach took over the leadership, the physiological declined sharply, and the psychological was almost completely forgotten. In short, the accepted outcome was that optics was a chapter of physics. If anyone dared to suggest that psychological elements formed part of optics, he was deemed to be little more than a philosophical dreamer searching for paradoxes and anomalies.

**58.** As presented at this time, optics was put together in the following way. Material bodies were aggregates of points or pointlike elements. From every element on the surface of these bodies rectilinear rays were emitted in all directions. Or rather, these rays were rectilinear so long as they traveled in a homogeneous medium. For if its index of refraction changed either abruptly or gradually, the direction of the rays also changed in accordance with the laws of reflection and refraction. Both these laws were subsumed under Fermat's principle that "the optical path $\int n \, ds$ between any two points of a ray is a minimum or a maximum."

When the change of direction occurred abruptly, reference was made to a reflecting or refracting "surface," and the reflection or refraction at this surface was investigated. The inquiry proceeded step by step, starting with a plane surface and then going on to a spherical surface, convex or concave. The examination of reflection generally stopped at this point and only rarely went on to analyze what happened when the rays were reflected by a second surface after having been reflected by a first, and then by a third, and so on.

In the case of refraction, transmission through a plane surface was studied. Then after a discussion of total reflection, the behavior of a plate with plane parallel faces was scrutinized, as well as that of a plate with plane but not parallel faces, namely, a *prism*. This involved tracing refraction through two successive plane surfaces. Thereafter refraction at a single spherical surface was determined, followed by that through two spherical surfaces, or *lenses*, and finally through systems of lenses and optical instruments.

**59.** This whole investigation in the nineteenth century proceeded essentially along geometrical lines.

In the study of the plane mirror a conventional little demonstration was regularly repeated in all the texts of whatever level. From a point $S$, rays are projected on a mirror $MM'$ (Fig. 11) and are thrown back in accordance with the law of reflection. The reflected pencil is divergent. But if each reflected ray is prolonged backward, it must pass through a point $I$, which is symmetrical to $S$ with respect to the plane of the mirror. Therefore, the bundle of reflected rays forms a cone with its vertex at $I$. The point $I$ found in this way was termed the *image* of the point object $S$; and since $I$ was the *point where the real rays intersected when prolonged backward,* it was described as a *virtual image.*

With regard to curved mirrors, it was shown that if the mirror had a small angular aperture, and a pencil of rays coming from a point *S* struck the mirror at angles of incidence differing little from

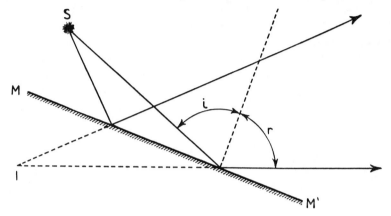

FIG. 11. Virtual image of a point object reflected in a plane mirror

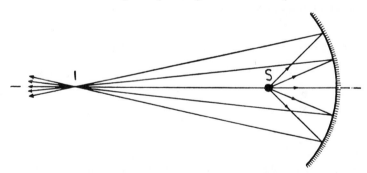

FIG. 12. Real image of a point object reflected in a concave mirror

the perpendiculars at the various points of incidence, the reflected rays again formed a cone with its vertex at a point *I*. The formation was not absolute and exact, however; but the approximation was closer, the smaller the mirror's angular aperture and the rays' angles of inclination.

Two cases could be distinguished. The pencil reflected by the mirror might be convergent, and then the point *I* was found on the reflected rays (Fig. 12). But the pencil might be divergent, and then the rays had to be prolonged backward behind the mirror in order to locate the point *I* (Fig. 13). In both cases the point *I* was named the *image* of the point object *S*. In the latter case, by obvious analogy with what happened in the plane mirror, the image was called virtual; in the former case, *real*.

Naturally, when the object consisted not of a single point *S*, but of an extended body, the foregoing reasoning was repeated point by point for every part of its surface that emitted rays. Thus point

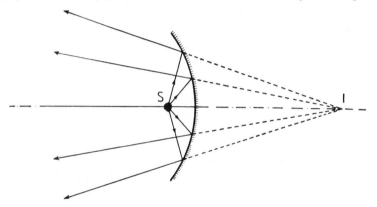

FIG. 13. Virtual image of a point object reflected in a concave mirror

by point its image was reconstructed. This might be real or virtual or also partly real and partly virtual.

**60.** These ideas were repeated without substantial variation in the treatment of refraction at one or more surfaces. In general, however, little study was given to the images produced by a single transparent surface, plane or spherical, and by a plate, whether with plane parallel faces or prismatic. Instead attention was concentrated on the behavior of lenses, which like curved mirrors yielded real and virtual images. In this connection Gauss's equation became classical:

$$\frac{1}{x} + \frac{1}{x'} = \frac{1}{f}$$

where *x* is the distance from the object to the mirror (or thin lens), *x'* is the distance from the image to the mirror (or thin lens), and

$f$ is a constant depending on the mirror (or thin lens) and given by the expression

$$\frac{1}{f} = \frac{2}{r}$$

in the case of the mirror, and

$$\frac{1}{f} = (n - 1)\left(\frac{1}{r_1} - \frac{1}{r_2}\right)$$

in the case of the thin lens.

There is no need to go into detail about matters familiar to all students of optics. In this direction the investigation attained remarkable precision and universality by formulating the concepts of *object-space* and *image-space*, and by showing in the most general manner that between the points of these spaces a one-to-one correspondence existed that was definable as a homography. Gauss's equations were the expression of this correspondence.

On these foundations the elementary theory of optical instruments was constructed. These instruments were optical systems capable of yielding real or virtual images. The most characteristic, like the telescope and microscope, gave virtual images. Others, like the camera and projector, furnished real images. Most of the known optical instruments were derived from these types.

**61.** For the purpose of improving the performance of these instruments, investigations were undertaken to maximize the effective aperture of the optical systems. It was of course discovered that when the aperture was increased, the rays emerging from the optical systems were not arranged in *cones* having their vertices in the *points* of the image, whether real or virtual, but instead gave rise to more or less complicated configurations. These were referred to as aberrations, since the rays did not conform to the pattern required by the simple theory.

Thus the study of *optical aberrations* began with the aim of classifying them, identifying their causes, and finding ways of diminishing them and, if possible, eliminating them. In this field too the mass of research was truly enormous and also highly beneficial.

The fundamental concept was the following. When the rays emitted by a point object and deviated by an optical system pass through a point, the image is perfect. Otherwise it is defective, with the defect becoming worse, the greater the aberration of the rays or their deviation from the direction they should have to give a perfect image.

Hence algebraic and trigonometric procedures and rules were worked out for ascertaining how to vary the geometrical characteristics of the optical system so as to reduce the aberration of the rays. The results were by no means insignificant.

**62.** In the foundations of optics the idea that "light consists of waves" became firmly established at the beginning of the nineteenth century. The marvelous fecundity of the wave theory, and the equally marvelous precision with which it accounted both conceptually and quantitatively for the phenomena of diffraction, interference, and polarization filled nineteenth-century physicists with such enthusiasm that they were convinced they had said the last word about "the nature of light." A veritable multitude of theoretical and experimental investigations ensued.

For some decades there was a curious promiscuity of rays and waves in the study of optical problems. The habit of using the geometrical model was so general that nobody thought it possible to examine a question of optics without talking about rays, even if later he had to end up by speaking about waves. But finally the entire range of known optical phenomena came to be expressed in terms of waves.

The luminous object thus became a body whose surface elements emitted spherical waves into the surrounding space. These waves kept their spherical (or plane) form so long as they were propagated in a homogeneous medium. But at a surface of separation between two media in which their velocity of propagation was different, the waves were deformed because the wave length varied proportionately with the velocity of propagation. The wave had to advance one wave length for each cycle of vibration completed by the source. Accordingly, when the wave front was traversing the medium with the lower velocity, it remained behind where it would have been in the medium with the higher velocity. This wave phenomenon explained the action of prisms and lenses.

When a divergent spherical wave emitted into the air by a point object *S* passes through a biconvex glass lens, it is deformed (Fig. 14). Its retardation is greater at the center of the lens than at the rim. Hence it becomes convergent, and may also be spherical with

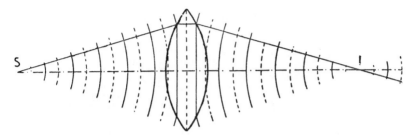

Fig. 14. Spherical wave deformed by passing through a converging lens

its center at a point *I*. Here it returns to its divergent state, as if *I* were a point object. Therefore *I* is the real image of the object *S*.

**63.** When this relation had been made clear, the rules governing the propagation of waves through diaphragms (such as the rim of the lens) together with the accompanying diffraction phenomena were applied to the image *I*. The conclusion was that a so-called spurious disk must be found at *I*. It is circular, if the rim of the lens is circular (Fig. 15). Its linear dimensions are greater, the longer the wave length of the radiation employed, and the smaller the angular aperture of the wave converging at *I*.

More commonly, this rule was expressed independently of the distance from the lens to the image. For the angular magnitude of the radius of the first dark ring of the disk, as seen from the optical center of the lens, is given by the equation

$$\gamma = \frac{1.22\lambda}{D}$$

where *D* is the effective diameter of the lens. This result of the wave theory of optical images underwent considerable development in two directions.

It became clear, in the first place, that entirely apart from any imperfection on the practical side, the image of a point is not a point but a disk of finite dimensions. This is usually referred to as **a** "spurious disk," although actually there is nothing spurious about it. It is always seen at the center of a spherical wave that has passed through a circular aperture. Since it occurs at the wave's center, it may properly be designated a *centric*, a term not pre-empted for any other purpose in optics. Because the centric is the perfect image of a

Fig. 15. Centric (diffraction figure at the center of curvature of a spherical wave which has passed through a diaphragm with a circular rim; sometimes called "Airy spurious disk")

point source, it was no longer possible to hope to obtain in the image a reproduction of all the details of the object when the latter was of appreciable size rather than a point source.

Consequently it was realized that every optical system, although ideally perfect, had a limited capacity to reproduce in the image the details of the object. For when two centrics approached each other, they could seem confused even if their centers were distinct (Fig. 16). Thus arose the concept of an optical system's *resolving power*, measured by "Lord Rayleigh's criterion," using the equation given above for $\gamma$. This amounted to saying that two centrics could not be distinguished as separate or resolved when the distance between their centers was less than the radius of the first dark ring of each of them.

In the second place, it was learned that when a wave emerging from an optical system was not spherical, there was no longer a center and therefore no centric either. Instead there were diffraction figures. These were more complicated, deformed, and spread out, the more the surface of the effective wave differed from a segment of a sphere. Thus the aberrations, which had previously been defined and studied geometrically, were treated as deformations of waves. This had highly important effects on the design of instruments. Among the most significant may be mentioned "the rule of a quarter of a wave length" and the resulting definition of "optical perfection." For when the waves emerging from an optical system were deformed less than $\lambda/4$, the images produced by them were diffraction figures differing so little from perfect centrics that they could practically be considered as such; hence that optical system yielded the maximum of which it was capable with those dimensions.

The results attained by technical optics in this direction were of the greatest practical value. The use of interferometers to study wave surfaces and the characteristics of optical systems was a decisive triumph.

Fig. 16. Merging of two centrics as they approach each other

**64.** Apart from the advance in wave studies, optics in the nineteenth century made good progress also in photometry and colorimetry despite difficulties and complications encountered at every step.

One group of investigations aimed at ascertaining how bodies reflect and transmit the visual radiation. For the most part these studies were successful. But serious obstacles impeded the definition of the fundamental concepts, photometric magnitudes, and related units. At one time only intensity was referred to, and white light; there was photometry in white light, and heterochromatic photometry. The "candle" was defined by taking a flame as a standard. But when it came to making measurements with a little care, it was discovered that the standard fluctuated. These measurements were made with visual photometers, which required of the observer's eye only a matching judgment.

When photoelectric cells and photographic methods were introduced, it was at once thought that it might be useful to substitute objective apparatus for the human eye in photometric measurements. Thus objective photometry was created, in contradistinction to subjective photometry, carried out by the eye.

At first the use of objective devices was limited, on account of their very low sensitivity. But as a result of their continuous and remarkable improvement, they were adopted more and more widely. The introduction of such sensitive and accurate means of detecting the visual radiation permitted photometric measurements to be carried to a precision previously unattainable. Consequently subjective photometry was almost completely eliminated from experimental procedure. At the same time the distinction between photometry in white light and heterochromatic photometry lost almost all its significance, at least conceptually.

The outcome of this long labor was the organization of the photometric magnitudes in the following group: quantity of light, flux, intensity, and luminance or brightness; to these, illuminance or illumination may be added. The corresponding units of measurement were defined, and also the methods of measurement. These methods are entirely objective; that is, they refer to an average eye, whose sensitive properties have been defined by international convention as the average of a considerable number of actual human eyes.

**65.** In the field of colorimetry the labor lasted longer and the organization came later. The correspondence between wave lengths and the colors of the spectrum induced physicists to conclude that the nature of color could be regarded as a problem already solved. But in the early nineteenth century it was pointed out that there are also nonspectral colors, not contained in the series of colors of the rainbow, and that the spectral colors have innumerable shades of varying purity.

In the course of these observations the trichromatic principle emerged. It held that any color may be obtained by mixing in the proper proportions three spectral (or even nonspectral) colors, called the "primary colors." Although this principle found wide application in the printing of colored illustrations, its development in the scientific field was very slow. Only in recent decades did this area too achieve an organization codified in international agreements.

At first the three primary colors were believed to have a physiological justification in the structure of the retina. But no matter how much research was done, this opinion was not confirmed. On the contrary, the tendency was steadily in the direction of a very interesting generalization. The three primary colors were not identified. It was necessary to conclude that the triad of primary colors was arbitrary in the sense that the three colors could be chosen at will (apart from the practicality of the choice). It was proved that what was essential in the triad was only the number three; that is, it was shown that colors are defined by three parameters, which may be the percentages of three fundamental colors, but may also be three other entities such as luminance, dominant wave length, and saturation.

Colorimetry accordingly undertook to represent the colors by means of triads of numbers, whose sum is equal to unity. Consequently a pair of co-ordinates is sufficient to identify a color, since the third number is unity minus the sum of the first two.

In practice it had to be acknowledged that the use of visual or subjective methods gave rise to very uncertain results. Hence objective methods were devised that employ photoelectric detectors of the radiation. The conclusions, however, are valid only for a conventional eye, whose sensitive and chromatic characteristics have been defined by international agreement.

**66.** After what was said in Chapter I, it should not surprise anybody if at this point I merely mention the infrared and ultraviolet radiations, discovered at the beginning of the nineteenth century; photography, established in common use about the middle of the century; spectroscopy, studied by physicists for many decades; the electromagnetic theory of radiation; investigations of the aether, forerunners of the theory of relativity; radiation emitted by a black body; the quantum theory; and other topics of this sort. This vast mass of research was incorporated by many persons more or less openly into the optics of the nineteenth and the first half of the twentieth centuries. They did so because they had too broad a conception of the meaning of the term "optics," which nobody undertook to define. But, as I tried to make clear in § 20, it is not reasonable to include all that material in optics. What I shall say hereafter will demonstrate even better the need for an explicit clarification of this question.

The reader who has glanced at the foregoing succinct summary (§§ 58–65) of the contents of optics in the past century and the first half of the present century will surely have noticed there the fundamental concepts that he learned at school from the time he was first introduced to the subject. He will have recognized there the ideas that were presented to him as "truths," confirmed, accepted, and beyond dispute. Yet the very fact that these "truths" have been so prominent in the preceding pages should arouse the suspicion that with regard to them there is still something to be said.

# CHAPTER III

# *The Foundations of the Science of Vision*

❖❖❖❖❖❖❖❖❖❖❖

**67.** In this Chapter, I propose to analyze visual optical phenomena, considering them in their physico-physiologico-psychological complexity, while trying to keep the hypotheses and philosophical assumptions introduced down to the smallest possible number.

The fundamental phenomenon is "vision," the act whereby an individual sees luminous and colored figures before him. When in a given situation nobody sees any such figures, we say "it's dark." If under certain conditions some persons do see such figures, and someone else does not, the latter is described as "blind." Because the blind do not enjoy the faculty of vision, students of that subject of course cannot take them into account. Hence all the people to whom reference will be made in the following pages will be nonblind.

To carry out our study, we must portray the mechanism of vision in some way. What I am now going to set forth summarizes over two thousand years of research. Yet it is so far removed from the ideas accepted by the general run of students that it should be presented, in my opinion, not as the scientific "truth" of the moment, but only as a working hypothesis, to be subjected to the trial of comparison with experience, in order to see how well it meets the test. After that, the appropriate conclusions will be drawn.

**68.** For vision to take place, there must be someone who sees, an observer. For our purposes, the observer consists of a living organism's mind. I must at once admit that I do not know what the human mind is. It is commonly said to be a function of the brain; and since a more precise determination of its nature is of no interest for our inquiry, I shall use this expression only for the sake of convenience, without letting it influence our reasoning.

Whatever the mind may be, it is closely related to the brain, the central organ into which flow impulses transmitted along the nerves connecting it with the peripheral sense organs. These nerve impulses, having been received in the brain, are analyzed by the mind, which draws conclusions from them.

So far as vision is concerned, the most important impulses are those that come from the eyes, along the optic nerves. Sometimes, however, impulses from another source intrude, as for example those derived from the muscles that control the rotation of the eyeball in its socket.

**69.** When vision takes place, there are present, in addition to an observer, also an observed object and radiant energy linking it with the observer. In a dream, object and energy may be absent, for in that state the observer has his eyes closed and beholds figures that seldom correspond to any object before him. There are likewise cases of persons who, while completely awake, see figures not seen by any other observer near them. This phenomenon of *hallucination* is considered pathological and abnormal. When the observer is not dreaming and not subject to hallucination, he sees by receiving energy signals that come from external objects in front of his eyes. The actual existence of these objects and signals cannot be proved absolutely, being one of the most perplexing problems of philosophy. Without entering into so thorny and delicate a discussion, I assume the material existence of the so-called "external world."

In keeping with the terminology used in physics today, I regard the external world as consisting of matter and energy. Matter is conceived as an assemblage of corpuscular units. Energy is thought of as something emitted by matter, and also absorbed by it, and propagated by the material body emitting it to the body absorbing it. The existence of this propagation has likewise not been proved absolutely, but is postulated.

**70.** The mechanism of this propagation has been and still is one of the unsolved problems of physics. At first a flow of material corpuscles was imagined; then, a series of undulations in a supposed elastic medium named the "aether"; later, a system of electromagnetic waves; finally, a stream of photons or tiny grains of energy. Some investigations, which study the phenomena exclusively from the geometrical point of view, take into consideration only the trajectories of the corpuscles or photons, or the perpendiculars to the undulations or waves (these trajectories or perpendiculars being the *rays*).

Each of these mechanical or geometrical models today is inadequate for a complete treatment of the known optical data. Although the wave and corpuscular concepts seem contradictory to us, they must be used simultaneously or alternatively for an exhaustive description of an entire optical phenomenon. To insist that only one of these ideas shall be employed necessitates restricting the discussion to partial phases of a phenomenon.

Only the corpuscular mechanism has hitherto succeeded in explaining the interactions between matter and energy, namely, the emission and absorption of the radiation. On the other hand, only the wave mechanism furnishes a useful picture of how the radiation is propagated. But every complete phenomenon consists of the emission of radiation by a material body, its propagation around the body, and its absorption by another body. Obviously, therefore, both mechanisms must be utilized, unless we wish to limit our investigation to only part of the phenomenon.

**71.** Accordingly we may say that for vision of the external world to take place (by contrast with dreams and hallucinations) there must be a body emitting radiation, which is propagated in the surrounding space, transmitted to an eye, and absorbed by it. A *stimulation* of the peripheral sense organ thus occurs, and impulses are generated which, by way of the optic nerve, carry to the brain and mind not only the news that the radiation has arrived at the eye but also the characteristics of the stimulus. Then the mind goes into action. It analyzes this "information," inferring therefrom whatever it can with regard to the form and position of the emitting body as well as its external properties. Having completed this operation, the mind portrays the conclusion with a model, a figure, an *effigy*. So too a sculptor, having received the description of an object by mail, studies it, analyzes it, and finally makes a clay model of it. The mind uses no clay, but constructs this nonmaterial effigy, which it then places outside the body at the distance deduced from the data received. Moreover, it endows the effigy with the form and colors that it utilizes to represent the incoming information.

Having finished this task, the mind confronts the luminous, colored effigy and says that it "sees the object" which emitted the radiation.

This conception of the visual mechanism may seem paradoxical and acrobatic, being far removed from the common mode of thought and the doctrine taught in the schools of the entire world. But I submit that it is the only version which can account for the optical phenomena known today. I intend to subject it to the harshest of criticism in the following pages, and the reader may judge for himself how it stands the crucial test of comparison with experience in the most diverse cases.

**72.** Now that the mechanism of vision has been presented in general outline, we must go into detail a bit in order to get a better understanding of the models that we shall have to use. The sequence is rather long: emitting body, emitted radiation, eye, nervous system, mind, effigy. Furthermore, most of the time we look, not at a source of waves in the true sense of the word, but at bodies that receive waves from a source, absorb them, and partly re-emit them in a modified form. Hence the sequence acquires another member: in addition to the emitting body, we shall have to deal with the re-emitting body too.

This sequence pertains to the simplest case, *direct vision*, in which the waves between the observed body and the eye traverse a homogeneous medium. But our investigation will have to be expanded to its greatest extent when treating the cases in which the course of the waves is interrupted by the modifying agents called "optical systems." The complete sequence, therefore, emerges as follows: emitting body, emitted radiation, body receiving the radiation and re-emitting it, re-emitted radiation, optical system, eye, nervous system, mind, and finally, effigy.

Were it necessary to mention everything known today about each link in this long chain, several volumes would be required. I shall, therefore, limit myself to recalling the ideas that are particularly pertinent to our inquiry.

**73.** The emission of radiation from bodies, for instance, is an enormous subject, which especially in the present century has been organized satisfactorily and splendidly developed. I shall, however, restrict myself to a few brief remarks.

The radiation with which we shall have to concern ourselves is emitted by the atoms of bodies, when these atoms are suitably excited. The emitted waves may have some particular wave lengths $\lambda$, or all the wave lengths included in a vast range of values. The series of these $\lambda$'s has acquired the name *spectrum*. If the $\lambda$'s are all those contained in an extensive region, the spectrum is termed *continuous;* otherwise it is labeled a *discontinuous* or *line* or *band spectrum*. These designations were suggested by the appearance of spectra observed through spectroscopes or photographed by spectrographs.

I wish to point out explicitly that in this discussion I systematically avoid using the adjective "luminous," so widespread in current scientific literature. For instance, a body that emits waves is generally referred to as a "luminous source," and spectra are frequently termed "luminous spectra." I urge the reader to shun expressions of this sort because implicit in them, even if they are used conventionally, lurks a dangerous thought that should be carefully scrutinized. In the interests of precision, a body that emits waves will here be called a "source of waves" or a "source of radiation." In the same way and for the same reason, I shall describe a body that receives waves, not as "illuminated," but as "irradiated."

**74.** The radiant energy emitted by these sources is measured by physical means, either nonselective detectors like thermoelectric piles, bolometers, and radiometers, or selective detectors like photoelectric cells. The measurement is expressed in terms of the joule, or its multiples or submultiples.

Instances are rare in which the really interesting feature is $E$, the total quantity of energy emitted by a source. If the emission lasts a long time with uniform flow, it is much more useful to know the power supplied, or energy emitted in a unit of time. This is called the *radiant flux F* and is measured in watts. The duration of emission being indicated by $t$, we have

$$F = \frac{E}{t}$$

This equation may be written in the form

$$F = \frac{dE}{dt}$$

when the emission occurs nonuniformly.

Just as it is more advantageous to know the emission's rate of flow in time rather than the total emission, so it is very often desirable to know also the distribution of the flux in the space surrounding the source, that is, in the various directions. Hence consider a point source, and divide the space around it into a number of elementary cones with their vertices in the source. A cone's solid angle, measured in steradians, is indicated by $d\omega$. The element $dF$ of flux per unit solid angle $d\omega$ defines $I$, the *intensity of radiation* in the direction of the axis of the elementary cone under consideration; in symbols,

$$I = \frac{dF}{d\omega} \qquad \text{or} \qquad I = \frac{F}{\omega}$$

the latter equation being applicable to the case of an emission uniform in all directions. The intensity of radiation is measured in watts per steradian.

If, however, the source is not small enough to be treated as a point, the concepts explained above hold for every element $d\sigma$ of its surface. To represent the distribution of the emission by the various elements of the source, consider the *radiance r*, defined as

$$r = \frac{dI}{d\sigma}$$

where $dI$ is the intensity emitted by the element $d\sigma$, $r$ being measured in watts/steradian m². 

When it is important to know, not the emission of each element in each direction, but the total emission of each element in the entire space around it, consider the *emittance* $\epsilon$, defined as

$$\epsilon = \frac{dF}{d\sigma}$$

and measured in watts/m².

Measurements of the flux, intensity, radiance, and emittance provide a useful description of the emission of radiant energy in time and space.

**75.** We come now to propagation, which requires a much more extensive examination. To avoid departing too far from our principal theme, the ideas of importance to us are gathered together in an Appendix at the end of this book.

We deal next with a body that receives radiation from a source and re-emits it. This too is a very intricate subject, and I shall limit myself to recalling only a few of its concepts. A material body may absorb incident radiation. This loses its wave structure and, becoming converted into other forms of energy, enters the body's atoms by a process the opposite of that by which it left the source's atoms. This absorption may give rise to complex phenomena like fluorescence and phosphorescence, consisting of the re-emission of radiation by the atoms excited by the incident radiation. In other cases the absorbed energy changes into heat, or more generally into internal energy. This chain of events does not concern us because, as a result of them, the irradiated body is transformed into a source, or the radiation is eliminated as such and therefore is no longer of any interest for the understanding of vision.

Highly interesting, on the contrary, are other instances in which the radiation, when it encounters material bodies, passes through them or is thrown back by them, while preserving its structure or wave length. A body that allows incident radiation, at least in part, to pass through it is termed *transparent* or *translucent*. A body that throws at least part of the incident radiation back is called a *specularly* or *diffusely reflecting* body.

**76.** These two phenomena of transmission and reflection must be considered from not only the geometrical but also the quantitative point of view. For example, in the case of reflection, we may

want to know what fraction of the incident radiation is thrown back
by the body receiving it.

A *polished* surface, whether plane or of some other form, behaves
like a reflecting optical system, and its behavior will be examined
from this point of view in §§ 307–312. If the polished surface is
assumed to be perfectly plane, so that as a consequence the reflected
beam follows the law of reflection strictly (that is, if the incident
wave is plane, the reflected wave also is plane) it remains to be

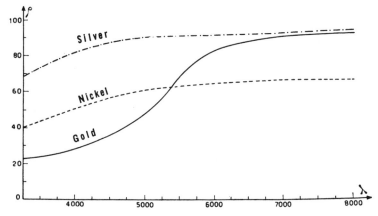

Fɪɢ. 17. Reflection coefficient ρ as a function of wave length λ, for silver,
nickel and gold

determined whether the reflected beam's intensity is equal to or less
than the incident beam's intensity.

This ratio varies with the specific characteristics of the material
composing the reflecting body, and may be expressed either numerically or graphically by indicating in numbers or curves the
*reflection coefficient* ρ, which is the ratio between the intensities of the
reflected and incident beams. This coefficient, being generally a
function of λ, may be plotted on a graph with λ as the abscissa
(Fig. 17).

In the case of transparent substances, the value of ρ is given by
Fresnel's well-known equations as a function of the refractive
indices of the media separated by the reflecting surface, and of the
incident wave's state of polarization.

If the reflecting surface is not polished but *diffusing*, even when the incident beam is projected in a single, well-defined direction, it is reflected in all directions. This phenomenon may be viewed as a whole by measuring the incident flux and reflected flux, thereby determining the reflection coefficient as the ratio of the second to the first of these fluxes. But it may also be required to know how the intensity in the reflected beam is distributed. For this purpose, this intensity is measured in various directions and represented by means of a *photometric solid*. This consists of a number of vectors, having a common origin in the incident beam's point of incidence and a length proportional to the intensity measured in the direction defined by the vector under consideration. Fig. 18 shows sections

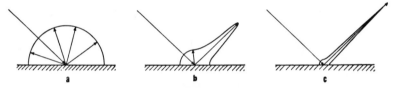

FIG. 18. Photometric solid of a perfect diffusing surface (*a*), of an ordinary body (*b*), and of an almost perfect reflector (*c*)

of the most characteristic photometric solids. The perfect diffusing surface (*a*) reflects equally in all directions; the ordinary diffusing surface (*b*) sends a good part of the radiation in the direction of the polished reflection, but also a not inconsiderable part in other directions; and finally the almost perfectly polished surface (*c*) reflects nearly the entire incident radiation in a single direction.

**77.** Similar concepts apply to bodies that are transparent to radiation, although geometrically the conditions are a little more complicated. The geometrical behavior of transparent bodies bounded by polished surfaces will be considered in §§ 313–322. Now we are concerned only with the quantitative aspect. For after saying that the radiation emerging from a body is less than the incident radiation, we face the problems of finding the reasons for this decrease and of indicating them in a useful way.

A body's *gross transmission coefficient* $\tau$ is defined as the ratio between the emergent flux and the incident flux. This is the ratio of the fluxes, it should be carefully noted, not of the intensities. These two ratios are not always equal. For a transparent body in general, even if it is extremely clear and has highly polished surfaces, behaves like an optical system that makes the radiation converge or diverge, thereby modifying the distribution of the intensity. Hence it is proper to refer to the ratio between the emergent and incident fluxes.

The amount $(1 - \tau)$ by which unity exceeds $\tau$ represents the fraction of the radiation that does not emerge from the body. This

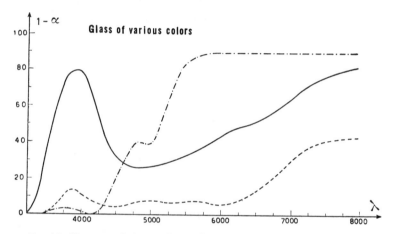

Fig. 19. Net transmission coefficient $(1 - \alpha)$ as a function of wave length $\lambda$, for glass of various colors

fraction is in part absorbed by the material composing the body, and in part reflected by its surfaces. As was indicated in § 76, the reflection coefficient $\rho$ gives the fraction of the incident radiation that is reflected. $1 - (\rho + \tau) = \alpha$ stands for the fraction of the incident radiation absorbed by the substance of the body, and $1 - \alpha$ is the *net transmission coefficient* of the given substance. This expression is a specific constant, which is a function of $\lambda$, and it too may be plotted on graphs having $\lambda$ as the abscissa (Fig. 19). To show more clearly the specific character of this magnitude, it should be made independent of the body's thickness, of which it is obviously an exponential function, by imagining the body to be divided into layers of unit thickness.

Bodies are diffusing either because they do not have polished surfaces or because the material of which they are made is turbid. With regard to them too we may speak of a gross transmission coefficient $\tau$, defined as above, and likewise of a net transmission coefficient $(1 - \alpha)$. But at the same time it may be necessary to resort also to the "photometric solid," consisting of a number of vectors originating in a point of the body's surface of emergence, oriented in all directions outside the body, and having a length, as measured on a certain scale, proportional to the intensity transmitted in every direction under consideration. Here too there are perfect diffusers, and those that are more or less efficient, down to the completely transparent body which, having received the radiation from a definite direction, transmits it in a single, equally definite direction.

**78.** To conclude this part of our review, let us recall that for the purpose of indicating how a source's radiation reaches a body under consideration, a magnitude has been defined that has not yet acquired a settled and generally accepted name, but that may with reason be called the *irradiance R*. This expresses the radiant flux impinging upon a unit surface of the receiving body. If $dF$ is the flux arriving at the surface $dS$, then

$$R = \frac{dF}{dS}$$

We should now go on to recall the fundamental ideas about the structure of optical systems and their effect on the radiation. But this topic will be examined in great detail in Chapters IV–VI.

**79.** We come then to the organ of vision. The human eye has been an object of study for thousands of years, and many volumes have been devoted to describing its anatomy and functions. Here too it is my intention not to go into minute particulars, but rather to summarize the essentials schematically.

The human visual apparatus consists of two balls, contained in sockets alongside the nose, and moved by special groups of muscles (Fig. 20). A section of a ball may be made (Fig. 21) by a plane passing through the axes of both eyes. This plane is sensibly horizontal, if the person whose eyes are under consideration holds his head erect. Each eye is a globe averaging 23 mm in diameter and bulging out in front.

Its external coating is a white, tough membrane, the *sclera*, whose anterior portion, the *cornea*, is transparent and protruding

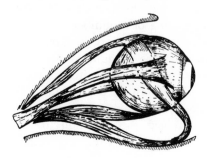

(as has already been mentioned). To the sclera's inner surface adheres another membrane, the *choroid*, which is dark brown in color and opaque. Attached to its interior is a third membrane, the *retina*, which for us is the most important.

Toward the front of the eyeball is found an interesting structure known as the *iris*, a diaphragm that is opaque except in

FIG. 20. Muscular system of the human eye

albinos. It is pierced in the center by an opening, the *pupil*, which is generally circular and variable in diameter. Along the junction of the iris with the sclera, where the latter is transformed into the cornea, lies a cluster of fibers, the *ciliary processes*. These regulate

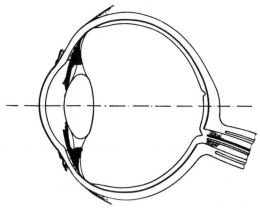

FIG. 21. Schematic section of the human eye

the shape of the *crystalline lens*, a strange gelatinous biconvex body located directly behind the iris.

The *anterior chamber* (the space between the cornea and the iris) and the *posterior chamber* (the space between the iris and the crystalline lens) are both filled with a perfectly transparent fluid, the *aqueous humor*, which is similar to water or rather to a very dilute salt solution. The space between the crystalline lens and the retina is filled with a jellylike fluid, the *vitreous humor*, which also is transparent. A third humor, the *lacrimal*, covers the external surface of the cornea like a very thin layer.

**80.** The function of the sclera, which forms a protective sheath around the whole delicate structure, is essentially mechanical. The choroid, an intricate maze of blood vessels, supplies nourishment. The retina, a highly complicated fabric of cells and nerve fibers, is specialized to receive the radiation.

The aqueous and vitreous humors, together with the cornea and crystalline lens, constitute a true optical system, which the radiation must traverse to reach the retina. Although the latter has been minutely investigated for a long time from both the anatomical and physiological points of view, it still remains a myriad of mysteries.

It is a membrane about 0.1 mm in thickness, consisting of various layers of nerve cells and ganglions, illustrated schematically in Fig. 22. Setting aside numerous complexities, which have not yet been fully described nor explained functionally, for our purposes we may regard the retina as a mosaic of many cells side by side. It has long been customary to subdivide these cells into two types, *rods* and *cones*, which occupy the second layer from the top in Fig. 22. Nearly all students of the subject believe these types to be endowed with different receptive properties, despite weighty objections that make the actual existence of such a duality highly doubtful. The number of cones in a human retina is computed by research workers as three to seven millions; the number of rods, as one hundred and ten to one hundred and thirty millions.

These cells, lying just beneath the retinal surface next to the choroid, are linked to nerve fibers (second layer from the bottom in Fig. 22) which then unite in a bundle, the *optic nerve*. That portion of it which leaves the sclera may be seen in Fig. 21. The number of fibers in the optic nerve has been variously calculated from thirty thousand as a minimum to more than 1.2 millions. Unquestionably there are several hundred thousand of them. Whatever the exact figure may be, it is much smaller than the number of rods and cones.

If these retinal units are estimated at one hundred millions, and the fibers of the optic nerve at one million, then as an average every fiber is the outlet for a hundred retinal units.

If this is the average, the linkage between the nerve fibers, on the one hand, and the rods and cones, on the other hand, is highly variable in different regions of the retina. Thus the *central fovea*, the part of the retina opposite the cornea and iris, contains only (or mainly) cones, and each cone is connected to one nerve fiber. In the *parafoveal region* around the fovea the percentage of rods rises, and goes still higher in the anterior portion of the retina, while the number of retinal units joined to the same nerve fiber likewise increases.

For the present, we may confine our discussion of the eye to these few brief remarks.

**81.** Even shorter will be our references to the nervous system that emerges from the eyes and goes to the brain. The two optic nerves enter the skull through two fissures at the back of the eye sockets. Then they unite in a strange junction, the *chiasm* (*C*, Fig. 23). As they leave it, they separate once more, but with an interchange of nerve fibers. Thus some of the fibers coming from the right eye form part, after the chiasm, of the nerve going to the left half of the brain, with the rest of the right eye's fibers in the nerve going to the right half of the brain; and the same holds true for the nerve coming from the left eye.

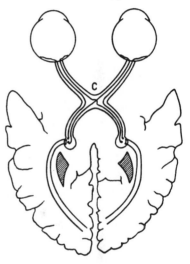

FIG. 23. Paths followed by optic nerve impulses in the central nervous system of man

---

FIG. 22. Schematic section of the human retina, showing the various layers. Above the top layer is the choroid, while below the bottom layer is the vitreous humor. Actual thickness of entire membrane is 0.1 mm (Polyak, *The Retina;* reproduced by kind permission of Mrs. Stephen Polyak).

In the brain the nerve fibers lead into various organs, which need not be mentioned now, until they reach the back of the head near the nape of the neck, where the cortical area is located that specializes in the phenomena of vision.

**82.** Our situation appears quite difficult. We have a sequence of elements and factors, and as we advance along it we see its complexity mount frighteningly and the extent of our knowledge diminish at the same time. Unfortunately, however, the goal toward which we are striving lies only at the end of this sequence.

It is easy to understand, and also to justify, the method followed by our predecessors. They cut the sequence at the point where it began to give them too many headaches, and they did not hesitate to declare that the process of vision came to a close at their cut. Some added, more accurately, "so far as we are concerned," meaning thereby that they were leaving it to the specialists in the other fields to pursue the investigation "so far as they were concerned." But even with this reservation their partial conclusions were not all correct; and the reservation itself was emphasized so little that finally it was overlooked and disappeared.

It is this kind of procedure on which I propose to turn my back. I intend instead to examine the phenomena of interest to us, taking proper account of the entire sequence outlined in § 72, even if to do so will not be easy and will often compel me to confess that certain links are still mysterious.

However, it is now the usual and established practice under such circumstances to introduce hypotheses where definite knowledge is lacking. It will indeed be very advantageous to enumerate explicitly the hypotheses we shall have to use for a complete description of the visual process. That enumeration will open a boundless field to anyone who desires to devote his research activity to it.

**83.** Let us begin with the simplest case, direct monocular vision of a point source. As was said in § 72, vision is called *direct* when the waves travel from the source to the eye without undergoing any deviation or deformation along the way. In this instance the sequence of factors is reduced to the minimum: source of waves, homogeneous surrounding medium (the air; if it were not homogeneous, it would be equivalent to an optical system, and vision would no longer be direct), eye, nervous system, brain, mind.

The source emits spherical waves with their center in the source. A segment of these waves encounters the cornea of an eye. At the outset of our investigation we must simplify matters as much as possible by taking only essential elements into consideration and postponing complications to a later time. Let us assume, for example, that the various transparent surfaces of the eye are spherical, with their centers of curvature in a straight line. Let us suppose also that the humors of the eye are homogeneous and that the crystalline lens, although admittedly not homogeneous, possesses symmetry of revolution with respect to that straight line.

This line is, therefore, an axis of symmetry of the whole eyeball, as thus simplified, and may be called the *optic axis*. Let us now postulate that the eye confronts the source of waves or is so oriented that its optic axis passes through the source, which of course is in front of the cornea. Thus the entire system, source + waves + eye, possesses symmetry of revolution around the optic axis.

In addition, let us take for granted that the distance from the source to the eye is very great, if not absolutely infinite. The usual and inevitable procedure, as is well known, is to start with geometrical and abstract expressions (such as "point, symmetry, infinite, spherical, plane"). The reasoning acquires pragmatic value, however, if these terms are replaced in later formulations by data that are more complex but realizable in practice. Hence there is no need to be deeply disturbed at present if we commence by considering a point source of plane waves propagated in a homogeneous medium, the source being at an infinite distance from an eye having symmetry of revolution around an axis passing through the source. This is a combination of ideal conditions useful in initiating an extraordinarily complicated course of reasoning, which must be approached with much caution and great patience.

**84.** These conditions are illustrated in a section made by passing a horizontal plane through *SOF*, the axis of symmetry (Fig. 24). The waves that are incident upon the cornea *O* are deformed by passing through it, as happens whenever plane waves encounter a spherical surface of separation between two media (§ 319). In our case, since the external medium is the air, which has a lower refractive index than that of the substances constituting the eye, the waves are made convergent. A further convergence is caused by the crystalline lens. Therefore, the waves that pass through the cornea, pupil, and humors converge at the back of the eye in the fovea *F*.

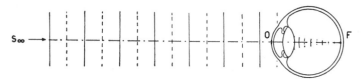

FIG. 24. Eye struck by a plane wave perpendicular to the optic axis

The conformation of most eyes is such that when the waves are made convergent, their center of curvature falls exactly on the retina. Hence a centric (§ 328) is formed there, the waves being absorbed by the matter present and ceasing to exist as radiant energy.

What happens in the retina as a result of the action of the incident radiation is still somewhat obscure. According to current views,

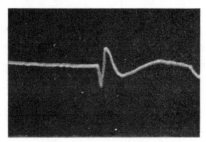

FIG. 25. Electroretinogram

the radiation is absorbed by the molecules and atoms of the substance contained in the retinal cells, displacing electrons there and provoking chemical reactions. Any substance especially disposed to react with minute quantities of radiant energy is in general termed *photosensitive*. Much important research work has been done to identify the photosensitive substance or substances active in the human eye, but the question cannot yet be regarded as settled. Because of this uncertainty, we shall resort to working hypotheses.

THE FOUNDATIONS OF THE SCIENCE OF VISION 85

That disturbances of an atomic and electronic nature occur in the retina, when affected by radiation, has been proved directly by detecting the resulting electric currents. When these are amplified by modern methods, they record *electroretinograms*, as they are called (Fig. 25).

The photochemical process must take place in the cells that make up the retinal mosaic. These are cones, because only cones are found at the center of the fovea. When they are stimulated by the radiation, nerve impulses are transmitted along the fibers linked to those cones, and are conveyed by these fibers along the optic nerve to the chiasm and from there to the cerebral cortex.

**85.** This topic is clearly neurophysiological, and unfortunately not much is known about it. The nature of the nerve impulse is still obscure. For a long time an important discussion has been going on between those physiologists who believe the impulse is electrical in nature and those who regard it as chemical. The latter thesis seems to be supported by the results of the most recent researches.

All physiologists appear to be in agreement about the following propositions, which are of fundamental importance for our study. All nerve impulses, whether they come from the eye, the ear, or the papillae of the skin, are identical in nature. In other words, not a single nerve impulse, regardless of the fact that it comes from a particular sense organ, contains within itself anything characteristic of the stimulating physical agent. Furthermore, according to the Adrian-Matthews *all-or-none law*, the impulses transmitted along a nerve are all equal to one another.

The mechanism of this transmission may be conceived as follows. When a sensitive nerve-ending is affected by a stimulus too weak to reach the threshold of sensitivity, stimulation does not occur and therefore no impulses are transmitted along the nerve. When the intensity of the stimulus crosses this limit, the process of transmission takes place. The energy carried by the stimulus is progressively absorbed, and when a given constant value is reached, a sort of flash occurs as the impulse leaves the stimulated nerve-ending for the opposite ending. At that instant- the sensitive ending remains refractory or inactive. Immediately thereafter it resumes absorbing the energy of the stimulus. No sooner has a quantity equal to the previous one been acquired than a second impulse is discharged, followed in like manner by a third, fourth, etc., until all the energy of the stimulus is exhausted.

**86.** According to this mechanism, the impulses propagated along a nerve fiber are all equal, as was indicated in § 85. Hence the transmission along the nerve can vary in only one way, the frequency with which the impulses follow one another. For the weaker the stimulus, the smaller its power (in the mechanical sense)

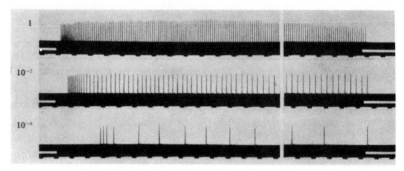

FIG. 26. Record of impulses discharged in a single optic nerve fiber of a king-crab (*Limulus*) in response to prolonged illumination of the crab's eye by three successive sources of different intensity, whose relative values are given at the left. Vertical lines indicate the impulses. Dashes below the horizontal rows register the passage of time in intervals of 1/5 second. End of white line in each row marks the beginning of the exposure to light (Hartline; reproduced by kind permission of the *Journal of the Optical Society of America*, 1940, *30*, 242).

and therefore the longer the time needed to accumulate the energy required for the discharge of the impulse to occur.

If the impulses were transmitted at a constant frequency, the information that could be forwarded by them would be limited to one thing, the intensity of the stimulus. The signals reaching the point of arrival would contain nothing from which any parameter could be extracted other than their frequency. Modern technique permits the detection of nerve impulses, and the results of research point to the conclusion that the mechanism must actually operate in this way in the lower animals (Fig. 26). In the higher animals (Fig. 27), although complete data have not yet been obtained, it is certain that the transmission is performed by a more complicated mechanism.

Transmission of nerve impulses at constant frequency would violate the general rule that Nature utilizes to the maximum the potentialities of the means at its disposal. For even by using a single nerve fiber it is possible to communicate more information, while still obeying the all-or-none law, if instead of a constant frequency a *modulated frequency* is employed.

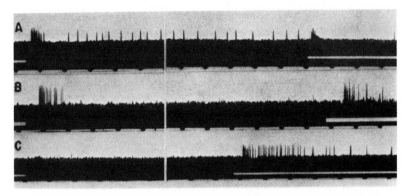

Fig. 27. Record of impulses discharged in single optic nerve fibers of a frog. In one type of fiber (row A) impulses are discharged regularly as long as the light shines. In another type (row B) impulses are discharged only when the light is turned on and again when it is turned off. In a third type (row C) the response occurs only when the light is turned off (Hartline; reproduced by kind permission of the *Journal of the Optical Society of America*, 1940, *30*, 244).

Radio successfully introduced transmission by frequency modulation, which is defined by three parameters: carrier frequency, modulated frequency, and index of modulation. Transmission by frequency modulation may be imagined as a periodic alteration of a transmission by constant frequency. The latter is, therefore, the carrier frequency. The frequency of the periodic alteration is the modulated frequency. Since the alteration consists of condensations and rarefactions of the impulses, the ratio of the minimum frequency to the maximum is taken as the index of modulation (Fig. 28). This mechanism can, therefore, transmit three types of information along a nerve fiber.

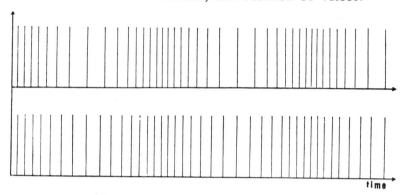

FIG. 28. Schematic diagram showing transmissions of impulses along a nerve fiber with equal carrier frequency and equal modulated frequency but with different indices of modulation

**87.** There are no experimental data to confirm a transmission of this type along the optic nerve. Nevertheless I shall consider the transmission of impulses by frequency modulation as a working hypothesis, which will serve us well in depicting the phenomena of interest to us.

Accordingly I shall suppose that the most complicated layer of the retina contains organs capable of responding to the energy of the stimulus and absorbing it, the result being a transmission of impulses by frequency modulation along a fiber of the optic nerve.

The admission of this hypothesis presents no special difficulties, since no extraordinarily complicated mechanisms are required for a transmission by frequency modulation. The modulated frequency may be regarded as the difference between two frequencies that are constant but of course different. For this effect to be achieved, it would be enough if the retina possessed two photosensitive substances of different sensitivity. But it is still much too early to make any assertions about this matter, while my hypothesis suffices for us to proceed with our analysis.

**88.** The wave radiation concentrated in a centric has, then, stimulated cones in the fovea. But how many cones?

Now we must make some numerical calculations. Let us recall that the wave lengths $\lambda$ of the visual radiations range between 0.4

and 0.8 $\mu$; the diameter $D$ of the pupil is 2 mm under *diurnal* conditions of radiation, and somewhat greater under *nocturnal* conditions; the refractive index $n$ of the vitreous humor is approximately 1.33; and the depth $f$ of the eyeball is 20 mm. The centric formed in these circumstances has, on the basis of the fundamental equations for the diffraction of waves (§ 328), a central disk of radius $r$, given by

$$r = \frac{1.22\,\dfrac{\lambda}{n}}{D}f = \frac{1.22 \times 6 \times 10^{-4} \times 20}{2 \times 1.33} = 5.5 \times 10^{-3}\,\text{mm} = 5.5\mu$$

Within the disk there is a maximum of energy at the center and a diminution along the radii, until zero is reached at the value of $r$ computed above. It is useless to take account of these details, however, because they correspond to optical perfection of the various ocular surfaces and media traversed by the radiation, and that perfection is never found in practice.

We may, therefore, limit ourselves to the conclusion that with a pupil 2 mm in diameter, the radiation emanating from a point source at an infinite distance is concentrated in a disk about $10\mu$ in diameter. Of the entire flux conveyed by the wave, more than 4/5 is concentrated in this disk, so that the surrounding rings may be neglected.

**89.** In the disk is found a small group of cones, perhaps four or five. They are stimulated and as a result, along the nerve fibers linked to them, trains of nerve impulses leave for the brain. Here the mind intervenes, analyzing the impulses received and trying to extract from them the maximum information about the source that emitted the waves.

When all this information has been deduced, it must be *represented*. For this purpose I attribute to the mind, as was indicated in § 71, the faculty of creating *effigies*, whose forms and colors depict precisely the characteristics inferred by the mind from the trains of impulses received by way of the optic nerve. I attribute to the mind the additional faculty of locating these effigies, thus created, at a suitable distance in the space before the eyes. When the mind has created the effigies and located them in front of the eyes, the ego says that *it sees the objects* of the external world.

To attribute these faculties to the mind is not a strange action without any foundation. The mind certainly enjoys such faculties, as is shown by its capacity to dream (§ 69). In a dream the ego beholds bright colored figures before it (and also hears voices and sounds) although it does not receive any radiant (or mechanical) stimulus from outside. There is no doubt, therefore, that those figures are created by the mind. In my view, then, it may be said that vision is like a dream built on the basis of information received by means of external stimuli and the peripheral organ of sight.

**90.** Let us carefully examine this process in detail, to see how the mind goes about constructing effigies on the basis of the information received by way of the optic nerves.

The fact that so small a number of cones at the back of the eye is affected means that the nerve impulses travel along a smaller (or at most, an equal) number of fibers of the optic nerve. Hence the mind infers that the dimensions of the source of waves are very small, or rather the smallest possible, because it never happens that the number of stimulated cones is less than in the present case. Accordingly the mind gives to the effigy the form of a *point* or, to be more precise, a disk of inappreciable diameter.

The arrival at the brain of impulses from certain fibers informs the mind that the affected cones are situated in the center of the retina, or rather of the fovea, on the eye's axis of symmetry (the existence of which was assumed in § 83). This information indicates to the mind that the source is on the optic axis and, therefore, supplies the mind with the *direction* in which it should locate the effigy out in front. The average person may not realize that there is an optic axis, and that the direction from which the waves arrive corresponds to the position of the stimulated cones in the retinal mosaic. But in § 93 we shall see how the mind proceeds to establish this correspondence.

**91.** To put the effigy in a definite place, the mind must determine how far away from the eye the source of waves is. This is the most difficult problem that the mind is called upon to solve. Let us leave it unanswered for the time being, because we shall have to discuss it at length in §§ 96–121.

Let us suppose for the moment that this determination too has been made. The effigy in the form of a "point" is thus located in the direction of the optic axis at the determined distance. But the effigy must have an aspect, for otherwise it is nothing. The mind must, so to speak, "depict" it by giving it a "luminosity" (which in technical terms is "intensity" or "brightness") and a "color." It is with these traits that the mind represents the energy characteristics of the train of impulses that it receives. The higher the carrier frequency, the livelier the brightness with which the effigy is endowed. The modulated frequency appears as color tone or hue, expressed by such words as "red, green, yellow," etc. The index of modulation corresponds to what is called "saturation."

In the nomenclature of colorimetry the liveliest colors, like those of the solar spectrum, are said to be *saturated*. Mixing a red with white produces a color of the same hue, but less lively, which is described as "less saturated." As more white is added, the red becomes less lively, until it looks "rose." As white continues to be added, it changes in the direction of "pale rose," and finally the original red hue is no longer felt. These gradations illustrate the concept of saturation.

**92.** When the mind, having created the effigy as a minute disk endowed with a certain intensity, hue, and saturation, has placed it in front of the eye on the axis at a certain distance, it says that *it sees a luminous point.*

As a matter of fact nearly everybody says that he *sees the luminous object* or *luminous source* or *"star."* Almost everybody is completely convinced that what he sees is the luminous source or star. Yet I must insist on reminding the reader that he should not permit himself to go along with this uncritical way of regarding the question. Even if he is not fully persuaded by my exposition, I ask him to follow the argument to the end, where he may draw his own conclusions.

From now on, to apply the term "luminous source" to the figure beheld in the circumstances under consideration is to conclude before thinking (§ 73). For the present, I must point out that there is a "source of waves," and a "luminous figure." To consider them distinct, for the purpose of discussing how the transition from the former to the latter occurs, prejudges nothing. Should we decide that they are identical, we could adopt the expression "see the luminous source." But it is obviously premature to do so in advance. For our final judgment may be favorable to distinguishing the two things.

**93.** Let us proceed with our analysis of the visual process by removing, one at a time, all the limitations imposed heretofore.

We may begin by eliminating the supposition that the point source of waves is on the axis of the eye, although still remaining at an infinite distance. Then plane waves reach the eye, but they are

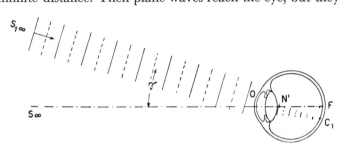

Fig. 29. Eye struck by a plane wave inclined to the optic axis

no longer perpendicular to the axis $SO$, as in Fig. 24. Now they travel along a direction $S_1O$, inclined at an angle $\gamma$ to the axis (Fig. 29).

In passing through the cornea and other media of the eye, these waves too are made convergent. As a first approximation we may say that they become spherical with their center on the retina, but at a point $C_1$ different from $F$, which in Fig. 24 is the center of the spherical waves coming from the extremity of the axis at infinity. If the source $S_1$ lies above the axis, the center $C_1$ is below it (§ 321).

Around the point $C_1$ there is again a centric. It will not be circular like the centric around $F$, but its dimensions will be of the same order of magnitude. It is not worth while to ascertain its precise dimensions, because these are made complicated by the slight irregularities found in every eye. The form of the centric around $C_1$ may be said to approach that around $F$ as the angle $\gamma$ diminishes.

Once more, then, a small group of retinal elements is stimulated. These are no longer only cones, but also rods, or even only rods. The difference between cones and rods, however, does not concern us at this moment. What matters is that the photosensitive substance around $C_1$ is modified, and consequently a train of nerve impulses of modulated frequency leaves for the chiasm and brain.

There the mind performs an analysis identical with the one already described in § 90, and the two sets of results differ in only one respect. This time the luminous colored point effigy is placed, not on the optic axis, but above it in a *direction* which forms an angle $\gamma$ with the axis and is determined by the *position* of the point $C_1$ on the retina.

**94.** Let $S$ and $S_1$, the two sources of waves, the former on the axis of the eye and the latter in a direction forming an angle $\gamma$ with the axis, be active at the same time. Then of course two trains of plane waves reach the eye. They intersect at an angle $\gamma$, but do not disturb each other. Hence two similar centrics are formed, one at $F$ and the other at $C_1$. In each of these areas a small group of retinal elements is stimulated. Along the nerve fibers linked to them, trains of nerve impulses of modulated frequency leave, simultaneously carrying to the brain information about the presence of two sources of waves in the space out front.

The mind, notified that there are two sources of waves, determines the dimensions, direction, distance, intensity, hue, and saturation of each. In conclusion it creates an effigy consisting of two luminous colored little disks, and places them at the indicated distance in the appropriate directions. It then sees "two stars."

Without needing any further explanation, we may assume that the sources of waves at an infinite distance number three, four, or more. For each of them we may repeat the analysis just made for the sources $S$ and $S_1$. Since the waves coming from one source do not disturb those emanating from the others, an equivalent number of small groups of sensitive retinal elements is stimulated. From each such group a train of nerve impulses leaves along the fibers linked to the group. The mind, having examined this train, is led to create a luminous colored point effigy. It then stations all these effigies before the eye in directions corresponding to the places where the respective retinal cells were affected. Thus the space in front of the eye is studded with a quantity of "stars" or luminous colored points, as happens when a person at night looks with one eye at the serene sky.

**95.** Let us now suppose that the plane waves emitted by one source $S$ on the axis and by another source $S_1$ reach the eye as in Fig. 29, but with the angle $\gamma$ very small (that is, below a limit which will be discussed in full detail in Chapter V). $F$ and $C_1$, the two centrics on the retina, may lie so close together that there are no unaffected cones between them. Then trains of nerve impulses travel to the brain along various adjacent fibers. When analyzed by the mind, these impulses inform it that the source of waves is not a point. Hence it creates an effigy that is no longer a tiny little disk, but is instead an extended figure with its major axis oriented like that of the group of stimulated retinal cones.

Let us consider next three, four, or more sources of waves, all very close to one another around the source $S$. Then many cells on the retina around $F$ are stimulated, and therefore trains of nerve impulses travel along a whole bundle of fibers toward the brain. There the mind draws the inference that the source of waves is not a point, but instead consists of numerous sources all clustered together so as to constitute a body of appreciable size. Consequently the mind creates an effigy with dimensions corresponding to the group of stimulated retinal elements. The mind then places this effigy before the eye at the determined distance so that every part of the effigy is in the direction corresponding to the affected area of the retina.

In addition, the mind analyzes each train of impulses that it receives along every fiber. It measures the carrier frequency, modulated frequency, and index of modulation. Accordingly, in creating the effigy, it endows each of the parts with the brightness, hue, and saturation corresponding to what occurred on the retina. Thus when a person at night looks at the sky with one eye, not only does he behold the stars but he may also see a luminous disk with brighter and darker areas. The observer says that he *sees the moon*.

**96.** We may now resume our consideration of the point source of waves in order to remove the restriction that it must be infinitely distant. Then its position $S$, which at first we shall assume to be on the axis, is defined either by its distance $x$ from the cornea $O$ or by its *vergence*, $\xi = 1/x$, with respect to the cornea (Fig. 30).

When the waves reach the cornea, they are not plane but spherical, the radius of curvature being $x$. Therefore, the wave refracted inside the eyeball has its center of curvature, not at $F$ as in Fig. 24, but beyond it at $C$. On the retina, then, the radiation is no longer concentrated in a centric, but instead fills a disk. This is larger, the nearer the point $S$ is to the cornea, or the smaller $x$ is and the bigger $\xi$ is. The general practice at present is to measure $x$ in meters, while $\xi$ is expressed in a unit called the *diopter*, abbreviated $D$.

To calculate how much $C$, the center of the refracted wave, is displaced from $F$ as the point $S$ approaches $O$, a very simple rule is

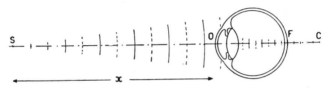

Fig. 30. Eye struck by a divergent wave with its center on the optic axis

available. For every $3D$ of the source's approach to the cornea, $C$ recedes 1 mm from the retina. Consequently, if $S$ is 0.33 m away from the cornea, the distance from the pupil to the retina being about 20 mm, the radiation on the retina will be spread over a small disk whose diameter is equal to 1/20 of the diameter of the pupil, or 0.1 mm when the pupil is 2 mm in diameter, the normal measurement in full daylight.

On the retina a disk of 0.1 mm diameter contains about 800 cones. From all of them nerve impulses now leave for the brain. The situation is exactly the same as it was in § 95, when the eye received the radiation from an object located very far away and subtending an appropriate angle $\gamma$. The mind reasons on the basis of the information which it receives. That information is the same now, when the object is a point $3D$ from the eye, as it was before, when the object was an extended body at a great distance. Hence the effigy created would have to be identical, unless the mind had a way of sensing the difference between the two cases.

**97.** For this purpose the crystalline lens intervenes. It enjoys the marvelous property of varying its own power by changing the curvature of its surfaces. How this happens is still being actively discussed, because the physiologists and anatomists have not yet succeeded in positively identifying the organs that produce this alteration in the shape of the lens. That the modification actually occurs is beyond question, since it has been confirmed by direct observation of *Purkinje's images*, reflected by the surfaces of the crystalline lens. It is commonly considered to result from contractions of the ciliary processes, fibers lying anterior to the *ora serrata* or

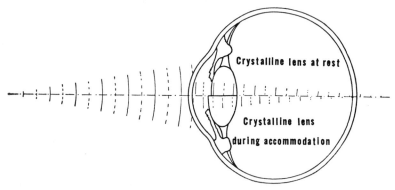

Fig. 31. Schematic diagram of the accommodation mechanism of the human eye

along the junction of the cornea with the sclera, where the choroid and retina both end (Fig. 31).

In any case, this function of the crystalline lens exists, and is called *accommodation*. It diminishes with advancing years, even in normal individuals, until it is completely lost about the age of seventy. For our present purposes it suffices to say that in normal young people accommodation takes place and is controlled by certain organs within the eye.

Let us now return to the situation in which all the retinal cones in a disk of 0.1 mm diameter are stimulated. Before creating an effigy, the mind orders the brain to have the strength of the crystalline lens increased. In the course of this operation the nerve fibers linked to the disk's peripheral cones stop transmitting their signals. Hence the mind infers that the source of waves is not very far off, but close by. Then it makes the crystalline lens continue the process of accommodation until the number of affected cones is reduced to the minimum. This happens when the power of the crystalline lens has augmented so much that the center of the concave wave inside the eye falls on the retina. There a centric is formed again, just as when the point source of waves was infinitely distant and the crystalline lens was passive (or unaccommodated). The effigy now fashioned by the mind is equal to that created under those conditions, as described in §§ 90–91. The present effigy has the intensity, hue, and saturation corresponding to the carrier frequency, modulated frequency, and index of modulation of the impulses transmitted along the optic nerve. It has the form of a "star" or little disk without apparent dimensions. It is placed along the optic axis, but no longer at a very great distance. Where then?

**98.** The mind now has a factor which indicates to it that the source of waves is nearby. That factor is the *effort of accommodation*, the effort that the crystalline lens must make to deform itself as much as is necessary to reduce the affected retinal area to a minimum. This effort is obviously greater, the nearer to the eye is $S$, the source of waves. The effort is commonly expressed in diopters. For example, an eye is said to have made an effort of accommodation of $n$ diopters when it deforms itself so as to reduce to a minimum the number of cones affected by a source of waves at a distance corresponding to $n$ diopters from the cornea.

Yet if the mind had to determine the distance $x$ from $S$ to the cornea in this way, it would not have sufficient data. When the accommodation is zero, $x$ is very large. If the accommodation increases, $x$ diminishes. But how much? The reply to this question can be given only by one of the other senses. It is touch that permits the eye to be calibrated by establishing a known ratio between the distance $x$ and the corresponding effort of accommodation.

When this ratio has been established (further details will be given in §§ 118 and 124) the conclusion may be drawn that in accommodation the eye has a telemetric mechanism, capable of measuring the distance $x$ from $S$, the source of waves, to the cornea. It may be said that thereby the problem of measuring this distance is solved, the problem I described as "the most difficult that the

mind is called upon to solve" (§ 91). But we must not deceive ourselves into thinking that our ship has reached port. The accommodation mechanism as a telemetric device is almost completely
ineffective for the eye and of virtually no practical importance for
ordinary vision. It becomes important only when it begins to be
inadequate. As a result of the aging of the mechanism with the
passing of the years, the accommodation effort of which a person
is capable no longer attains $4D$, and therewith the defect known as
*presbyopia* commences.

**99.** To realize the ineffectiveness of accommodation as a telemetric device, a few calculations are enough. It was pointed out in
§ 96 that when a source is $3D$ from the eye, the center of the wave
refracted in the eye lies 1 mm beyond the retina, as indicated in
Fig. 30, and hence the stimulated cones number about 800. But if
the source is $1D$ from the eye, the number of cones involved falls
below 100. If the source moves as far as $0.5D$ away from the cornea,
the number declines to about 20; and when the source reaches
$0.25D$, the number drops right down to 4 or 5 cones.

In other words, when the source of waves is 4 m from the eye,
the spherical waves incident upon the cornea are refracted so that
they stimulate the same number of cones in the retina as when the
source is at any greater distance whatever. Therefore accommodation never occurs when the source of waves is located at any point
4 or more meters from the eye.

This fact, which is of considerable practical importance especially in the testing of vision, is usually expressed by saying that
"for the eye, infinity begins at 4 m" (in the case of monocular
vision). This amounts to saying that the eye does not sense $1/4 \, D$
with the mechanism of accommodation. This result too is of considerable practical importance because it defines the tolerance for
optometric measurements and corrective eyeglasses, and therefore
rests on a very broad experimental basis.

**100.** This tolerance holds good for accommodation at least up
to $10D$ (because thereafter it becomes larger). Hence it indicates the
indeterminacy with which even an eye calibrated by long experience can gauge the distance of a source of waves by using this

mechanism alone. In Table I the first column gives the distance of the source of waves from the cornea; the second column, the indeterminacy in absolute units; and the third column, the indeterminacy as a percentage of the distance.

TABLE I

### INDETERMINACY OF DISTANCES ESTIMATED BY ACCOMMODATION

| Distance from the source to the cornea (in meters) | Indeterminacy (in meters) | Percentage |
|:---:|:---:|:---:|
| 0.1 | 0.0025 | 2.5 |
| 0.5 | 0.063 | 12.6 |
| 1.0 | 0.26 | 26 |
| 2.0 | 1.0 | 50 |
| 4.0 | ∞ | ∞ |

As a telemetric mechanism, accommodation is evidently not very effective. Nor is the slight result that it can accomplish always utilized, as we shall often see hereafter. Experiments have been performed for the purpose of determining the distance at which a point source of visual waves is seen by an observer using only one eye in completely dark surroundings. These tests have revealed the mind's utter uncertainty in this situation, provided that every precaution is taken to prevent information from reaching it in any other way.

For the mind does proceed to the termination of its task. It creates the "luminous point" effigy, which it must then place somewhere. The direction in which it can locate the effigy is well defined (the axis of the eye, in the present instance) and as for the effigy's distance, the mind does the best it can. When objective information is lacking or is vague, the mind supplies the deficiency on its own initiative. In the end the observer *sees* a star at a certain distance, and vows he sees it at that distance. Most of the time the source is much nearer or much farther away, and the observer does not believe or know that seeing the star at the distance at which he sees it is only the product of his own mind's initiative.

**101.** To resume our systematic exposition, we may now remove the restriction that the point source of waves, while remaining at a finite distance, must be on the optic axis. When it is outside the axis (Fig. 32), we may repeat what was said in § 93 about an infinitely distant source. But once again accommodation of the

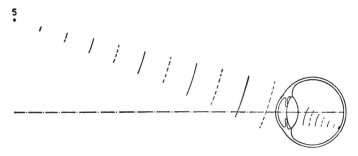

Fig. 32. Eye struck by a divergent wave with its center not on the axis

crystalline lens is invoked for the purpose of reducing to a minimum the group of affected cones, even if they are no longer in the center of the fovea but more peripheral.

Consider next two sources of waves $S$ and $S_1$, active at the same time and at the same distance from the cornea (Fig. 33). The source $S$ is on the axis, while the source $S_1$ is in a direction inclined to the

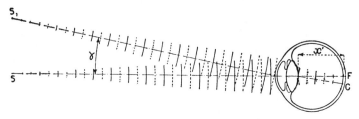

Fig. 33. Eye struck by two waves, one with its center on the axis, and the other with its center not on the axis

axis at an angle $\gamma$. When the crystalline lens is suitably accommodated, two small groups of elements will be affected on the retina, one at $F$ and the other at $C$. From the corresponding nerve fibers, trains of impulses will be transmitted to the brain. From them the mind will draw the data for creating two effigies (or a double effigy, which is the same thing) consisting of two "stars," each having a certain intensity, hue, and saturation. One will be located along the axis at the distance deemed proper by the mind. The other will be placed in the direction corresponding to the point $C$ on the retina, and therefore inclined at angle $\gamma$ to the axis. This star will be stationed at the same distance from the eye as the other star, should the mind be of that opinion.

If the sources of waves number three, four, etc., the effigies created by the mind also become three, four, etc. Each is endowed with the form (a star), intensity, hue, and saturation as well as location deduced by the mind from the information that it receives from the individual trains of nerve impulses transmitted by the fibers linked to the stimulated retinal units.

**102.** Thus when anyone beholds the starry firmament at night (under those conditions it does not matter whether he uses both his eyes) he sees all the stars as though imbedded in a vault, the so-called *celestial sphere*. In estimating the height of the vault, he might speak of hundreds of meters or a few thousand. The luminous points seen by the ordinary observer obviously do not coincide with the sources of waves, for those sources consist of the heavenly bodies. They differ enormously in their distances from us (the planets of the solar system being some hundred million kilometers from the earth, whereas the extragalactic nebulae are at least a million light-years away) and in any case they are all vastly more remote than the celestial sphere appears to be.

Experimental confirmation of the visual mechanism which I am expounding may be postponed for the time being. I have inserted this brief reference ahead of its proper place for the following reason. Should the assertion that the mind arbitrarily integrates the information received by way of the optic nerve be regarded by any reader as too daring, he may at once ascertain that there are clear cases in which this really happens. But we shall see better examples later on in Chapter IV.

**103.** If we now go on to take up the case in which the point sources of waves are very numerous and close together, constituting what is called a body, it is plain how our reasoning may proceed.

Every one of these sources sends waves to the eye. Since the waves of one do not interfere with those of another, each may be considered independently of the rest. Then an equivalent number of groups of cones will be stimulated on the retina and will form one continuous zone. From all these cones nerve impulses will leave along the entire bundle of fibers linked to them, the train of impulses along any one fiber being as a rule different from that transmitted along the other fibers.

Hence information flows to the brain in incredible quantities. It is analyzed by the mind, which not only constructs the effigy element by element, endowing each element with brightness, hue, and saturation, but also gives to the aggregate the form and position corresponding to the aggregate of the information received. This is a task which, if translated into numbers, acquires astonishing proportions, considering the speed with which it is completed. Only recently has anything like it been built—the complicated but marvelous electronic calculating machines.

For when anyone looks at an object, he receives waves emitted by its individual atoms. When he says that he sees it, he has already analyzed all the nerve impulses produced by all those waves, and has created the corresponding effigy. Looking at an object is generally believed to be the same as seeing it. This is not true. Between the two operations lies all that wonderful and intense work. We shall have to discuss it at length before obtaining a sufficiently clear conception of it.

**104.** Returning to the conditions of Fig. 30, let us suppose the distance $x$ between the source of waves and the cornea to be so small that the eye cannot accommodate adequately, or the number of diopters $\xi = 1/x$ to exceed that within the power of the crystalline lens. Then the wave inside the eye, although convergent, has its center beyond the retina. The number of retinal cones affected exceeds the minimum, which is found when the center of the wave lies exactly on the retina. The nerve fibers involved are more than one, and along each of them trains of impulses are transmitted at modulated frequency. The mind thus receives the same information as if the crystalline lens had accommodated adequately for sources of waves so numerous and close together as to form an extended object. The mind has no basis for a different decision, and therefore ends its work by constructing an extended effigy corresponding to the number of affected cones. The effigy's brightness is less than the created star would have had, could the crystalline lens have accommodated adequately. But because the modulation, as a first approximation, is equal for all the cones, the effigy's color will be uniform, and its saturation also. Hence, confronted by a point source of waves, the observer sees a luminous disk of appreciable dimensions.

Consider next two point sources of waves, both under the conditions described above, that is, at distances from the eye such that the crystalline lens cannot accommodate adequately. For each source we shall have to repeat the same reasoning. Consequently at the end of its work the mind will create, not two luminous stars, but two luminous little disks, whose centers occupy the place that would have been occupied by the two stars had the crystalline lens accommodated adequately.

Accordingly, if the eye confronts a group of sources of waves that constitutes a solid object at so small a distance that the crystalline lens does not command accommodation enough to reduce the affected retinal area to a minimum, the result is very complicated. To every elementary point of the object corresponds a retinal disk, the cones within which are stimulated by the waves emitted by that element. Since the individual elements are adjacent to one another, the *centers* of the retinal disks are adjacent too. Hence the disks partially overlap. Therefore each of the cones toward the interior of the affected retinal zone will receive a stronger stimulus, the resultant of partial stimuli belonging to various disks. But the peripheral cones (or rods) will receive stimuli that diminish in strength the nearer the cones (or rods) are to the boundary of the affected retinal zone.

**105.** In conclusion, the effigy created by the mind will be a broader and more confused figure than would have existed had the crystalline lens accommodated adequately. This contrast may be confirmed at once by bringing a finger to a distance of 4 or 5 cm from the eye. Instead of the bright figure rich in fine details, which is seen when the finger is as far from the eye as is necessary for the crystalline lens to accommodate properly, a confused figure with indefinite boundaries is seen, and hence people usually say that what they see is *blurred*.

Such a statement recalls how absurdly many persons reason who do not want to be persuaded that the figure seen, what I have termed "the effigy created by the observer's mind," is distinct from the material object, which I have called "the aggregate of the point sources of waves." These persons, convinced that the figure seen is always the object, speak of a "clear object" when it is seen well, and a "blurred object" when it is seen badly. But they fail to notice that such descriptions contradict their thesis. For the finger or object is always the same, whether it is placed near the eye or far away. Yet what the observer sees changes perceptibly and progressively from one position to the other. How can two things be deemed identical when one of them remains constant while the other varies?

A similar case occurs when two objects are aligned with the eye at distances so far apart that their vergences with respect to the eye differ by more than $1/4$ of a diopter, or better still by several diopters. A familiar instance is a pair of rifle sights, the rear sight being $1D$ and the front sight 4 or $5D$ from the eye. If the crystalline lens accommodates for $1D$, the rear sight is seen clearly but the front sight is blurred. If the front sight is seen clearly, that means the crystalline lens is accommodating for 4 or $5D$, and therefore the rear sight is blurred.

**106.** We may now eliminate another hypothesis previously adopted for the purpose of simplification. According to this supposition, the entire optical system of the eye was free from imperfections; that is, the media traversed were assumed to be homogeneous, and the surfaces of the cornea and crystalline lens to be spherical. No such conditions are found in any actual eye, as I remarked in § 88. Disregarding for the present certain gross structural deformities known as *ametropias*, we may turn our attention to the interesting case of *irradiation*, as it is called.

When the source of waves has the form of a point, inhomogeneities in the various humors, and above all in the crystalline lens, make the radiation concentrate on the retina, not in a centric with its center at the point $F$ (as in Fig. 24) but in an irregular zone. This consists of a very intense central nucleus and numerous offshoots branching out from the nucleus in many directions. Hence an extremely restricted group of cones is strongly excited, while the adjacent cones are stimulated along the radial offshoots.

Consequently the effigy created by the mind is not merely a point or little disk of inappreciable dimensions, but around it there are many "bristles" corresponding to the offshoots. What emerges is the familiar radiated "star," which is seen by all normal eyes when they look at a point source of waves, like the stars in the sky or even electric bulbs very far away. It is these appearances that constitute irradiation. This too is an eminently clear and common case wherein what is seen is so different from the material body that emits the waves. Yet also in this instance it is curious how rarely we find anybody mentioning the discrepancy.

For even if doubts may arise concerning what is around the celestial stars, which are so remote and inaccessible (I do not now desire to make use of the fact that those rays or bristles are never seen around the stars in the telescope), it is easy to ascertain, by an inspection at close range, how an electric bulb is made. Anyone undertaking to account for the phenomenon of irradiation must admit that when he looks at an electric bulb from afar so that it seems like a point, he sees it surrounded by an immense crown of rays, whereas from nearby he sees nothing at all around it.

To conclude from these facts that the crown of rays has no material existence requires but little reasoning. Hence it requires but little further reasoning to infer that when the crown is beheld, it must be regarded as having been created by the observer. This deduction may be readily confirmed by noticing that every eye, even of the same observer, sees a different crown around the same source of waves.

**107.** To summarize our analysis thus far, we may say that the mind constructs bright colored effigies on the basis of information received from the optic nerves as a consequence of stimulations produced in the retina by external agents.

This conclusion must not be construed, however, in precise and absolute terms. We may say, more fully, that the effigies are constructed on the basis of *all* the information that the mind receives in any way whatever, and also on the basis of information received in the past and stored in the *memory*. Sometimes it happens that the information received by the optical route is incomplete; then, to terminate its task, the mind combines this information with that coming from other sources, present and past, as has already been indicated in some cases (§§ 98, 100, 102). Not infrequently it happens that the information arriving by the optical route contradicts that derived from the other senses or preserved in the memory. The mind then finds itself in the difficult situation of being obliged to decide between two or more contradictory reports. How does it do so?

Naturally the answer cannot be categorical and general. Each individual behaves in his own way, according to the intelligence and critical spirit with which he is endowed, and not seldom on the basis of his imagination and cultural background. This is behavior with purely psychological characteristics, but it is inseparable from the study of optical systems, which this book has undertaken to examine.

**108.** To take account of this behavior, we should push a bit further our analysis of the means that are at the disposal of an observer for the purpose of obtaining information about the position of the sources sending visual waves toward him.

Assume that $S$, a point source of waves, is in front of a single eye at distance $x$ (Fig. 34). Consider $S$ to be moving very rapidly while $x$ remains constant; in other words, suppose the source of waves to describe an arc of a circle having its center in the eye. What happens on the retina is obvious. When the source of waves is at $S$, a small group of cones is stimulated at $F$, the group being reduced to a minimum by appropriate accommodation of the crystalline lens. When the source is displaced toward $S_1$, the cones at $F$ cease to be affected, while the adjacent cones in the direction of $C_1$ are stimulated. Then these too cease to be excited, while the cones adjacent to them are in turn aroused, and so on.

Thus the brain receives impulses coming in succession from different and adjacent retinal elements. Hence the mind infers that the source is always to be found in different directions corresponding to the position of the affected elements. Consequently, in accordance with the shift in the retinal impression, it creates an effigy in the form of a moving star.

The process of impressing the retinal units lasts a certain amount of time (of the order of a tenth of a second). Hence it may be that while the elements at $C_1$ are being stimulated, those at $F$ have not yet ceased to emit nerve impulses, and this of course holds true even

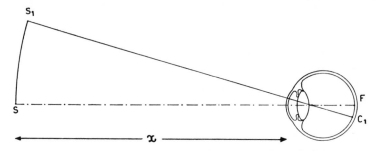

FIG. 34. Eye confronting a point source which moves transversely with respect to the axis

more for all the retinal elements between $F$ and $C_1$. Informed that the brain is receiving impulses from all the fibers linked to the elements between $F$ and $C_1$, the mind creates an effigy in the form of a luminous line between $S$ and $S_1$. As the motion continues, this line shortens on the side toward $S$ and lengthens on the side toward $S_1$.

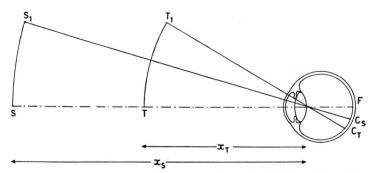

FIG. 35. Eye confronting two moving point sources at different distances

This is what happens when a person sees a falling star on a clear night.

**109.** If $S$ and $T$ are two point sources of waves at distances $x_S$ and $x_T$ from the eye (Fig. 35) and both move with equal speed, obviously the effects on the retina are not equal. When $S$ reaches $S_1$, $T$ has arrived at $T_1$. At the same time the impression, which was originally at $F$ for both sources, as shown by the broken line in Fig. 35, is felt at $C_S$ for the first source, whereas it has attained $C_T$ for the second. Therefore the retinal trace $FC_T$ exceeds $FC_S$ in length, exactly as many times as the distance $x_S$ surpasses $x_T$. Hence to the effigy corresponding to $T$ the mind attributes a greater velocity than to the effigy corresponding to $S$. In more familiar language this relation is stated by saying that when bodies move with equal velocity but at unequal distances from the eye, the more remote *seem* to move more slowly.

All this estimating of motions is done by the mind with care, because it constitutes a very powerful means of gauging the distance from the eye to the sources of visual waves.

**110.** Consider two point sources, like $S$ and $T$ in Fig. 36, aligned with an eye so that their impressions on the retina take

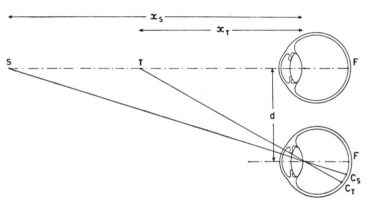

Fig. 36. Temporal parallax in monocular vision of two point sources at different distances from the eye

place at the same time in $F$. Then move the eye a distance $d$ so that the axis remains sensibly parallel to itself. The retinal impressions now due to $S$ and $T$ are felt at $C_S$ and $C_T$ respectively. The distance between $C_S$ and $C_T$ is greater, the more $x_T$ differs from $x_S$; or rather, when $d$ is small in comparison with these values of $x$ (as happens in the interesting cases) $FC_T$ exceeds $FC_S$ as many times as $x_S$ surpasses $x_T$.

Consider next two point sources, like $S$ and $T$, which are so far away that an observer using only one eye is not in a position to estimate their distance. Hence he cannot even perceive whether they are at different distances from the eye. But it is possible by a suitable displacement to feel at once that there are two distinct sources, that $S$ is farther away than $T$, and that it is so many times more distant. During the motion of the eye the impression on the retina is displaced. This displacement is analogous to what occurred in the case illustrated in Fig. 35. Now, however, the observer does not infer that the two sources are in motion, because the mind *knows* that the eye has moved. Hence it deduces that the sources of the waves are stationary but at different distances, and locates the effigies accordingly.

Should either or both of the sources $S$ and $T$ by some fortuitous coincidence execute an independent movement at the very same moment in which the eye is displaced, serious complications would ensue. The mind would be fooled. Who knows what conclusions it would draw from an examination of the impulses arriving at the brain, and what localizations it would select?

Deception is generally not easy, because the mind makes the eye carry out control movements. By a close comparison of the resulting variations in the retinal impressions, the mind almost always succeeds in solving the puzzle, placing the effigies reasonably close to the sources of the waves and endowing them with appropriate velocities.

This mechanism for studying the geometrical situation of sources of waves is known as *temporal parallax*. "Parallax" is the angle subtended at each of the sources by the eye's displacement $d$. This parallax is called "temporal" because it varies in time.

To become convinced of this mechanism's power, we need only perform experiments with a single eye, compelling it to observe through a fixed pupil and then permitting it to be displaced. In the former case, it is much easier to commit gross errors in locating the effigies. Of course these experiments must be done with proper precautions for preventing the mind from utilizing other sources of information. These other sources will be discussed in §§ 116–121.

**111.** I have referred briefly to temporal parallax because it offers us a convenient bridge for passing over to the mechanism of *binocular vision.* Heretofore we have always considered *monocular vision,* or vision with only one eye. But the normal individual has two eyes, whose centers are from 55 to 65 mm apart, 60 mm being regarded as the average. This is the so-called *interpupillary distance* (*ID* in Fig. 37).

To every stimulated element of the retina, like $C_1$ in Fig. 29, there corresponds a direction, like $N'S_1$, along which the source of

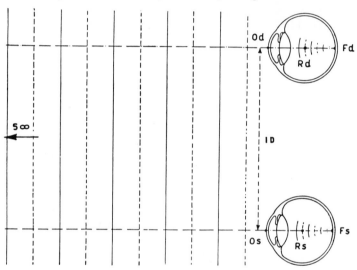

Fig. 37. Binocular vision of an infinitely distant point

waves must lie. This relation gives to the individual equipped with two normal eyes a very powerful telemetric means of determining the distance of the source from the eyes.

The retinas of the two eyes being sensibly symmetrical, a *correspondence* exists between the elements of one and those of the other. The mechanism of this correspondence is not yet entirely clear, although the chiasm is obviously involved. In any case the correspondence is felt by the mind.

Let $O_d$ designate the right eye, and $O_s$ the left eye (Fig. 37). With the sole aim of simplifying our statements, let us assume that the observer stands erect and looks straight ahead. Then the axes of his two eyes define a horizontal plane, which we shall suppose to coincide with the plane of the paper in our Figures.

For our purposes, each eye can rotate around a vertical axis, which cuts the optic axis in a point known as the *center of rotation* ($R_d$ for the right eye, $R_s$ for the left eye), which is close to the geometrical center of the eyeball.

**112.** Consider now a point source $S_\infty$ of visual waves, which is located in the horizontal plane determined by the axes of the two eyes, at a distance so great that it may be regarded as infinite. When the observer looks at this source, he directs the axes of his eyes at it. Hence the waves in the vicinity of the eyes may be treated as plane and perpendicular to the axes. These in turn, being directed toward a very distant point, may be deemed parallel.

Two segments of waves enter the eyes and, both crystalline lenses being relaxed, form two centrics at the foveas $F_d$ and $F_s$, because the situation illustrated in Fig. 24 is reproduced in each eye. $F_d$ and $F_s$ are to be considered corresponding points.

Then the brain receives at the same time nerve impulses from the fibers linked to the right eye and from those linked to the left eye. Both groups of impulses bring the same information for the mind. Therefore the conclusion is the placing of a point effigy having the form of a star in the direction of both optic axes at a very great distance.

**113.** Now move $S$, the source of waves, to a distance $x$ from the base line $R_dR_s$ (Fig. 38), this distance being less than a limit that will be determined in § 115. This time the waves reaching the eyes are sensibly spherical. The wave segments entering the two pupils, by proper accommodation of the crystalline lenses, have their centers at two points, one of which is $C_d$ to the right of $F_d$ in the right eye, while the other is $C_s$ to the left of $F_s$ in the left eye.

If the eyes remained stationary, the impulses coming from the right eye would impel the mind to infer that there is a source of waves to the left at a distance defined by the crystalline lens' effort of accommodation. At the same time the impulses arriving from the left eye would lead to the conclusion that there is a source of waves to the right. The mind would thus be induced to create two equal

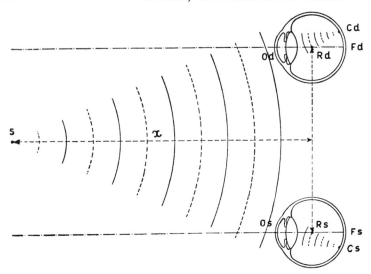

FIG. 38. Binocular vision of a point at a finite distance, with the eyes set for distant vision

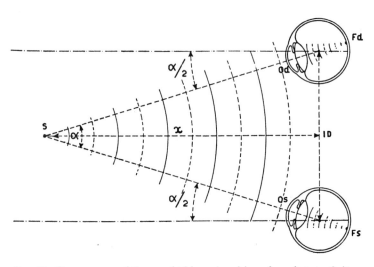

FIG. 39. Convergence of the eyes in binocular vision of a point at a finite distance

effigies, one to the left and the other to the right, in the horizontal plane of the axes. This well-known phenomenon of *diplopia* is ordinarily called "seeing double."

But the eyes generally do not remain stationary. They are rotated so that their axes pass through the source $S$ (Fig. 39). When that happens, the stimulated cones are again those at $F_d$ and $F_s$. Then the nerve impulses incite the mind to create a single point effigy where the two axes intersect. During the rotation of the eyes the observer watches the two effigies, created previously when the axes were parallel, approach each other until he sees them become one. This is the process known as *fusion*.

**114.** It is especially noticeable when two sources of waves are active at the same time, one close to the eyes (say, 30 cm) and the other farther away (a few meters). In performing such an experiment, point sources need not be used because every solid object may be considered an aggregate of point sources (§ 103). For example, hold a pencil 30 cm in front of the eyes, and at the same time look at a stick parallel to the pencil and 2 m away. If the observer fuses the effigies of the stick so as to see a single figure, he sees two pencils, one to the right and the other to the left, and *he sees nothing where the material pencil is*. If he rotates his eyes so as to fuse the two effigies of the pencil into only one (and then people say that they *see the pencil*) he sees two sticks, one to the right and the other to the left, and he sees nothing in the place where he previously saw a single stick.

Experiments of this kind, which could be multiplied, for instance by forcing the eyes (or only one of them) to rotate by pressure from a finger, show quite clearly that the effigy seen is of a piece with the eyeball and not with the material source of waves. Although people see such phenomena very frequently, they completely ignore them, strangely enough, and never ask what those double figures are that they see and how they ever become the *object* when they fuse.

**115.** Coming back to our analysis of the phenomenon illustrated in Fig. 39, we may note that to fuse the two effigies of the source $S$, the two eyes must each rotate through an angle $\alpha/2$, equal to half of the angle $\alpha$ which the interpupillary distance $ID$ subtends at $S$; that is,

$$\frac{\alpha}{2} = \frac{ID}{2x} = \frac{ID}{2}\,\xi$$

where $\xi$ is the vergence of the source $S$ with respect to the eyes. The angle $\alpha$ is called the *parallax* of $S$ with reference to the observer, and the rotation of the eyes is termed the *convergence* in binocular vision.

Convergence is accomplished by the action of the appropriate muscles that control the position of the eyeballs in the sockets (Fig. 20). Hence the relevant information is furnished to the mind by an agency other than the optic nerves. This agency is certainly more effective than accommodation. For now, in essence, the observer solves a triangle with a base of 60 mm, whereas when he looks with only one eye, the base is the diameter of the pupil, usually 2 mm, and therefore 1/30 as large. But obviously the mechanism of convergence also has a limit, whose order of magnitude it is worth while to determine.

Differentiating the previous equation, we have

$$d\xi = \frac{d\alpha}{ID}$$

where $d\alpha$ represents the smallest change in the parallax which the binocular system can perceive, and $d\xi$ is the resulting uncertainty in the determination of the vergence $\xi$ of the source of waves. The value of $d\alpha$ must be ascertained by experiment, and quite different results have been obtained. Observers highly skilled in continuous stereoscopic measurements show a minimum as low as 12'' (or approximately $6 \times 10^{-5}$ radians) but in untrained persons it may be considered as approximately 2' ($= 6 \times 10^{-4}$ radians). Then we have in the former case

$$d\xi = \frac{6 \times 10^{-5}}{6 \times 10^{-2}} = 10^{-3}D$$

and in the latter case

$$d\xi = \frac{6 \times 10^{-4}}{6 \times 10^{-2}} = 10^{-2}D$$

In the former case the distance of the source $S$ is determined within one thousandth of a diopter; in the latter case, within one hundredth. If we accept an average value of $0.005D$, we may construct the following Table.

TABLE II

INDETERMINACY OF DISTANCES ESTIMATED
BY CONVERGENCE

| Distance from the eye to the source (in meters) | Indeterminacy (in meters) | Percentage |
|:---:|:---:|:---:|
| 1 | 0.005 | 0.5 |
| 2 | 0.020 | 1.0 |
| 3 | 0.045 | 1.5 |
| 4 | 0.080 | 2.0 |
| 5 | 0.125 | 2.5 |
| 7 | 0.245 | 3.5 |
| 10 | 0.50 | 5.0 |
| 15 | 1.13 | 7.5 |
| 20 | 2.0 | 10 |
| 25 | 3.1 | 12 |
| 33 | 5.6 | 17 |
| 50 | 12.7 | 25 |
| 70 | 25 | 36 |
| 100 | 53 | 53 |
| 150 | 130 | 83 |
| 200 | ∞ | ∞ |

It is evident that beyond 200 m the binocular mechanism becomes completely ineffective. To repeat the expression adopted for accommodation (§ 99), we may say that "for binocular vision, infinity begins at 200 m." Even around 100 m, however, this mechanism is so inaccurate that its practical value may be put at zero. Those who have to rely on the estimates of ordinary observers using binocular vision regard it as useless beyond 25 m.

**116.** Yet in practice every observer constantly sees any number of figures at distances more remote than this limit. Seeing them there means, as was indicated in § 89, that the effigies have been placed there. Hence the observer had some way to determine the distance of the sources of the waves arriving at his eyes. Consequently there must be telemetric mechanisms in addition to those already discussed.

Anyone who gets to the bottom of this subject comes to the conclusion that such mechanisms exist, but are predominantly psychological in nature. In other words, the reports reaching the mind by way of the eyes are carefully analyzed and *integrated* with a mass of other information, some of it derived from geometry and perspective, but most of it drawn from the memory, that is, from previous related experiences.

There is a big difference between what is seen by a person looking at a landscape for the first time and by someone who has lived there for years. A statement of this sort will sound queer to many readers, who are thoroughly convinced that the figures they see are neither more nor less than the actual objects, which therefore cannot be seen otherwise than where they really are and as they really are. To shake so deeply rooted and so long established a conviction would require a protracted and detailed discussion with experimental demonstrations. But to avoid tarrying too long over this topic, let us examine its main features and concentrate our attention on the most important points. Above all I urge the reader to renounce beliefs that are *blind*, that is, uncritical and unconfirmed. To do so will suffice, because in his daily observation he encounters innumerable cases in which, confronted by the very objects themselves, he finds himself the author of changeable effigies, which are constantly being improved as a result of new data reaching his mind.

How often it happens that a new panorama appears limited and close to us, but then as time passes and we have explored and traversed it, it is seen to be spacious and vast! How often at the seashore a small boat seemed a few hundred yards away, but later was revealed to be a big, distant steamer, which looked unattainable when we tried to reach it in a rowboat! An evil spirit seemed to be moving it away from us while we were rowing toward it, and what had been anticipated as a pleasant little ride turned out to be an interminable, painful labor. So too in the mountains how often have massive peaks in the distance been viewed by novices as hills easy to climb! Instances of this sort could be strung out endlessly. But the person involved barely notices them with a momentary feeling of surprise and then promptly forgets them, because he is dominated by blind faith in the false proposition that the figures seen are the real objects.

**117.** The creation of the great variegated effigy that constitutes the *apparent external world* before the eyes of every observer is an extraordinarily complex problem. Beyond 100 m this effigy is constructed for the most part on the basis of comparison with the dimensions of familiar objects like men, animals, trees, and houses. Suppose, for example, that in the retinas of the eyes the group of cones excited by the waves emanating from the points of a house is much smaller than the group affected by the waves coming from a man. Then the mind infers that the house must be much farther away than the man, and it places the two effigies accordingly.

But after an examination of the impulses arriving from the cones stimulated by the waves that were emitted by the house, the mind may conclude that the house is smaller than the man. For instance, by extensive use of temporal parallax and considerable displacement of the eyes, the observer may become convinced that the house is nearer than the man. Then the mind may decide that it is not a real house but a model, and therefore place it nearer and make it smaller.

Illustrations of this kind of reasoning are available in inexhaustible abundance. Consequently I shall limit myself to a few standard examples. What matters most of all is to show how intricate and unsubmissive to general rules is the production of the effigy representing the apparent external world. This is a proposition of which we shall have to make comprehensive use hereafter.

**118.** Let us now return to the situation illustrated in Fig. 33. From our discussion at that time it is clear that $FC$ on the retina may be expressed as the product $\gamma x'$, $\gamma$ being the angular distance between the two sources $S$ and $S_1$, while $x'$ is the distance from the retina to $N'$, the eye's *second nodal point*.

Since this is a fixed point in every eye, $x'$ is a constant. This is the reason why in §§ 90 and 93 I said that to every *position* on the retina a definite *direction* corresponds. Consider, for instance, two objects $SS_1$ and $TT_1$, the former being exactly double the latter but at a distance $x_S$ from the eye also double $x_T$ (Fig. 40). Then both $SS_1$

and $TT_1$ subtend the same angle $\gamma$ at the eye $O$. Both $S$ and $T$, being on the optic axis, emit waves which, after being made convergent by the eye, have their center at $F$. $S_1$ and $T_1$, being also aligned with $O$, propagate waves which, having been made convergent by the eye, have their center at $C$. Therefore, whether the linear object in front of the eye is $SS_1$ or $TT_1$, the same retinal units are affected, namely, those between $F$ and $C$.

The only difference between the two cases is that a greater effort of accommodation is needed for the object $TT_1$. If $x_T$ is more than 4 m, however, there is not even this difference, because accommodation is no longer involved.

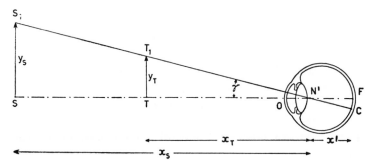

Fig. 40. Two different objects subtending the same angle at the eye

Then we must ask the following question. Since in both cases the retinal impression is the same, and therefore identical nerve impulses reach the brain, how does the mind decide whether it ought to create the effigy $SS_1$ or $TT_1$ or some other between these two or even beyond $SS_1$, provided always that it subtends the same angle $\gamma$?

In substance we should be inclined to answer that the problem has no solution. The only clearly defined element in it is the angular magnitude $\gamma$, which is necessarily and quite precisely connected with the affected retinal zone $FC$. As for the rest, the observer must try to determine the distance of the object by stretching his arm out or walking over to the object and touching it, if possible. This isn't always possible, as in the case of the heavenly bodies.

**119.** They provide a clear illustration of the expedient to which the mind resorts: it generally chooses the most economical solution.

Consider the sun and the moon, both seen from the earth within the same angle $\gamma$, about 32′. This equality of their apparent diameters is demonstrated in total eclipses of the sun, when it is covered by the moon almost exactly. But the sun is much bigger than the moon, its radius being nearly 500 times larger. Therefore its distance too is that much larger. Unfortunately our organism has no means of measuring these distances and consequently cannot perceive the difference between them, however enormous it may be. So the mind creates two effigies that are equal (apart from their brightness) and places them the same distance away. This is very much less than the moon's actual distance (300,000 km), not to mention the sun's (150,000,000 km).

Those who are convinced that when gazing at the heavens, they see the true sun and the true moon, are evidently far from the truth.

**120.** Even more peculiar is the familiar phenomenon known as the *flattening of the celestial sphere*. If this were truly spherical with its center in the observer, the sun's disk and the moon's would always be equal, whether overhead or near the horizon. But this is the case only for very few observers. Generally both disks, when near the horizon, seem much bigger, about two or three times bigger, than when they are overhead. If the angle $\gamma$ subtended by the disks in both positions is measured with a goniometer, the result is regularly the usual 32′, or at most the vertical diameter is found to be ever so slightly *smaller* when the disks are at the horizon.

Why then do they look bigger there? The answer is based on the reasoning illustrated in Fig. 40. When the sun is on the horizon, the effigy is placed two or three times farther away than when the sun is overhead, and the same holds true for the moon. The reason for this is that in the vertical direction there are no points of reference. The effigies of the stars are customarily placed behind the clouds, and these are judged to be half a kilometer up, more or less. In the horizontal plane, on the other hand, all the objects in the landscape induce us to conceive it as much vaster. Hence the celestial sphere is imagined as flattened, much less high than wide.

**121.** The connection between an object's angular magnitude $\gamma$ and its distance $x$ from the eye is utilized by the mind most of the time the other way around. Instead of determining $x$ and then creating an effigy of such dimensions that it will subtend the angle $\gamma$, *if the dimensions of the object are already known*, the mind places it at the distance required for it to subtend the angle $\gamma$. A landscape in which many familiar objects are scattered about lends itself to a fairly faithful reconstruction in the following manner. Unknown objects are placed with reference to the known by means of parallaxes or other expedients. I shall merely mention, among others, estimating the clearness, cloudiness, and color of the various figures. For sometimes the mind even takes into consideration that more distant figures look darker and more bluish than those nearby.

Yet the situation is not so simple. Considerable advantage would be derived from the equation

$$\frac{y_S}{x_S} = \frac{FC}{x'}$$

if this were the relation of the dimensions $y_S$ of the effigy, its distance $x_S$ from the eye, and the size of the retinal trace $FC$. But in reality this rule too is set aside by more basic principles operating in the mind. When objects are observed less than 15 or 20 m from the eye, the rule can no longer be trusted. Knowing that objects do not alter their dimensions even if they recede from the eye or approach it, the mind creates effigies of unchanging dimensions, however much their angular magnitude changes.

Thus a person sees his own hand always of the same size, whether he holds it 20 cm from his eyes or extends it to 30, 50 or 60 cm, where its angular magnitude decreases to one third, and the retinal trace $FC$ decreases to one third. This is true not merely for the hand, because the same thing happens with the furniture in the room, the objects on the table, etc. The problem of explaining how the mind proceeds in creating the effigy that constitutes the apparent external world is highly complex, as I said in § 117 and now repeat. I trust that the reader is convinced.

**122.** In summarizing the concepts reviewed somewhat schematically in this Chapter, I should like to emphasize that the figure that a seeing individual beholds before himself, in a field containing sources of visual waves, is a creation of his own mind. This arrives at the figure by combining *all* the factors at its disposal. A part of these factors consists of the modulated nerve impulses that reach the brain by way of the optic nerves. Another part comprises information that is muscular in nature, being connected with the mechanisms of accommodation, convergence, and temporal parallax. Still another part is composed of previous knowledge of the observed objects and past experience with them, the memory acting as the storehouse. Another part, finally, is contributed by the imagination and initiative of the mind itself.

This whole mass of factors is analyzed with a precision and rapidity that are marvelous. The end product is a judgment or series of judgments by the mind, which may reason soundly or unsoundly. If it blunders too often, it is adjudged abnormal and what it says it sees is called a *hallucination*. Sometimes it succumbs to errors that are committed more or less by everybody under the same conditions, and then it is said to be subject to an *illusion*. At other times it makes a mistake that it notices; it corrects the created effigy and says nothing, even if a bit surprised at first. Most of the time it fashions an effigy that it does not subject to any criticism or control. For the effigy serves its needs well enough, and it is satisfied. It believes that the figure it sees is the actual reality, forgetting that it has itself created the figure and given it brightness and color.

**123.** To complete this picture, sketchy as it may be, I should like to mention how it has been confirmed by extremely interesting observations of several dozen individuals who, although blind from birth, acquired sight when they were old enough to describe intelligently the sensations they experienced.

In their condition, called *congenital cataract*, the crystalline lens is opaque at birth. Hence the waves emanating from external sources are absorbed before they can affect the retina. But after the opaque crystalline lens is removed, the waves are able to reach the sensitive region. A suitable converging lens is put in front of the eye in order to secure effects comparable to those of normal eyes. The lens is designed in conjunction with the cornea to deform the waves so as to make their center fall on the retina.

If the diagnosis of those who are born blind indicates this opacity, the crystalline lens is usually destroyed early in life, when it is very difficult to interpret the expressions by which the patients try to convey what they see. But in some cases, particularly in the past century, the operation was performed on persons over twenty who were thoroughly conscious of their behavior while blind and capable of depicting with accuracy what they saw as soon as they acquired the ability to feel the visual waves.

**124.** Their behavior has been reported with many fascinating details in voluminous works. Here it will be enough to recall that the first time the visual waves reached the retinas of these patients, they felt a sensation of brightness and color. But it was undefined in form and position, and they could not recognize any of the surrounding objects, even in their essential features, although these were well known to them by touch, as they are to the blind. In order to recognize an object, even one as familiar as a watch, they had to touch it with their hands so as to establish a correlation between the figure seen (that is, the created effigy) and the form of the object as known by touch. They were obliged, as it were, to "calibrate" their retinas so as to discover the rule for locating a given effigy created in response to the stimulation of a certain retinal zone.

For this to be done with nearly normal speed and accuracy, about two months of training were generally needed. During this period of time those who had just acquired vision may be said to have been learning to see, or learning to use their eyes, exactly as others learn to use any instrument whatsoever. In that interval they made the strangest and most unexpected mistakes. Naturally the worst were in the perception of depth, or the estimate of distance. Not infrequently they judged a pencil a few feet away from their eyes to be equal to a belfry many hundreds of yards away. But as their skill improved by constantly comparing the figures seen with the object's form and position as determined by touch, the correspondence between the two sets of data gradually grew better until it attained the level found in normal persons.

These too of course have equivalent training, which they receive automatically as infants. It is common knowledge among parents that little babies want to *touch everything*, and they often scold them for insisting and try to make them give up this *bad habit*. Actually it is not a bad habit, but a necessity. It is the natural way in which they learn to know the external world, learn to establish relations between retinal stimuli and tactile sensations, learn to locate the effigies with precision, in short, learn to see.

**125.** Finally I should also point out that the information reaching the mind is not always consistent. In particular, the reports coming to it along the optical pathway do not always coincide with those arriving by another route. Such conflict will be of special interest to us in Chapter IV, and we have already glanced at some examples of it in binocular vision.

For instance, as was indicated in § 114, if a person looks at a pencil held vertically in his hand about a foot away from his eyes and at the same time looks at an upright stick a few yards beyond, there are two possibilities: either the observer looks at the stick and sees two pencils, or he looks at the pencil and sees two sticks. Yet he knows by the sense of touch that there is only one material pencil. Hence when he looks at the stick, from his right eye a report reaches his mind that there is a pencil to the left of the stick; from his left eye, that there is a pencil to the right of the stick; and from his hand holding it, that there is only one pencil, directly in front of the stick.

But we need not resort to binocular vision in order to land in this sort of quandary. As we saw in § 105, anyone aiming a rifle equipped with a front and a rear sight must solve the problem posed by conflicting information. For if he looks at the rear sight some eight inches from his eye, he sees it clearly defined, but at the same time he sees the front sight with its outline blurred. If with the same eye he looks at the front sight about a yard away from his eye, he sees the rear sight with its outline blurred.

Instances of this kind happen very frequently to all observers. Yet those who heed them are few indeed. For most of the time the mind dismisses the question without calling attention to it. Once precedence has been granted to a certain batch of reports, the mind ignores the others as if they had not reached it.

Anyone who does concentrate on the situation is amazed how often it occurs, and how complicated and even annoying the effects are when, after they have been noticed, they can no longer be disregarded.

# CHAPTER IV

# Vision by Means of Optical Systems

◇◇◇◇◇◇◇◇◇◇◇◇

**126.** In Chapter III we examined the rules governing the mechanism of direct vision. Let us now go on to study vision that is not direct. This is the case when visual waves emitted by sources, instead of reaching the eyes directly, are deviated or deformed along the way. The agencies producing this deviation or deformation are in general called *optical systems*. We shall begin with the simplest of them all, the *plane mirror*, which is a bright flat surface usually consisting of a thin layer of metal adhering to a sheet of glass.

It is the function of a plane mirror to deviate waves without deforming them (§§ 308–309). The principle according to which the deviation occurs is the familiar and ancient "law of reflection." For our present purpose it should be noted that if spherical waves emanating from a point source $S$ impinge upon a plane mirror $MM$, they are deviated, while keeping their spherical form, as if they had been emitted by a source $S'$, symmetrical to $S$ with respect to the plane of the mirror (Fig. 41). So far as physics is concerned, there is nothing behind the mirror that affects the waves.

If the reflected waves reach an eye, they enter it and are rendered convergent, just as if they had arrived directly. By proper accommodation of the crystalline lens the center of the convergent waves is carried to the retina. By appropriate movements of the muscles of the socket, the center of the waves may be brought to the center of the fovea. Here recommences that whole sequence of operations that we analyzed in Chapter III, and it goes on until an effigy is constructed by the mind of the observer and placed before the eye. Where is it stationed?

If the center of the convergent waves has been made to fall on the center of the fovea, the eye has turned its axis perpendicular to the waves, that is, in the direction of $S'$. We should therefore be

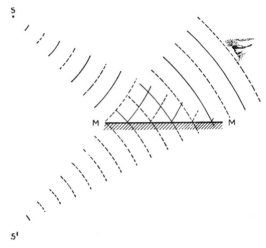

FIG. 41. Vision by means of a plane mirror

inclined to conclude at once that the point effigy must unquestionably be located on the straight line connecting the vertex of the cornea with $S'$.

**127.** This conclusion is not absolutely necessary. But what it states actually happens because, when looking at or through optical systems, most observers proceed in the manner to which they are accustomed through constant practice of direct vision. Hence we may say that the point effigy created by the mind of the observer will be located along the straight line joining the corneal vertex to $S'$. At what distance will it be placed?

It is clear from the lengthy discussion in §§ 96–121 that the answer to this question is very difficult and uncertain. The mind of the observer finds that it must solve the usual telemetric problem. It can take advantage of accommodation, temporal parallax, and convergence (for the reflected waves may affect both of the observer's eyes). But when it wants to resort to information arriving by another route, complications begin.

Suppose there is no information of this kind at all. To eliminate it is hard, but not impossible. For instance, the following experimental conditions may be arranged. The point source of waves is enclosed in a black opaque box. Its one opening is turned toward the mirror so that only this receives the waves. The mirror is highly polished and entirely free of defects and dust or other matter on its surface. Its edges are black and covered by black screens in a black and dark background.

Under such circumstances (which are not easy to achieve as completely as is required) the observer's eyes receive only the waves reflected by the mirror, and receive them exactly as if they had been emitted by the point $S'$, symmetrical to the source $S$ with respect to the mirror. His mind then functions precisely as in the case of direct vision. If the distance from his eyes to the point $S'$ is about a meter or, better still, less than a meter, the effigy in the form of a luminous point may be placed very close to $S'$.

But if the observer, *having seen the luminous point at $S'$*, wished to make sure of its existence, for example by touching it, obviously he would not succeed. If he stretched his arm out, he would bump it against the mirror. He would have to be satisfied with what he *saw*, in other words, accept his own creature and its position as valid and regard it as *true*.

**128.** But when the distance to be determined goes up to several meters or, worse still, several dozen meters or even more, the telemetric problem becomes both physically and physiologically so indeterminate that the luminous point effigy is placed with great uncertainty and with results that differ markedly from one case to the next and from one observer to the next.

An experiment performed on a group of observers (who were advanced students of optics) led to the following findings. A point source, carefully concealed in an opaque container, projected its waves on an optically perfect plane mirror 16 cm in diameter and 18 m away. The mirror was set up at the end of a completely darkened corridor. Behind the mirror was the (blackened) outside wall of the building. The observer was put near the source. The point symmetrical to it with respect to the mirror was therefore 36 m away from the observer. He was free from all restrictions; in other words, he could employ both his eyes and also make full use of temporal parallax.

Yet all the observers knew that even though the mirror was completely invisible, the building ended behind it. This was information that the mind of every observer unconsciously was unable to set aside. For it would have been absurd to imagine a star that was outside the building yet seen through its thick opaque external wall.

The effect was general in all the observers. Instead of locating the luminous star 36 m away, at the center of the waves reaching their eyes, they placed it *on the mirror* 18 m away, where they knew the corridor ended. The optical reports to the contrary were so weak and uncertain that they failed to oust those derived from the memory and the environment.

**129.** To understand the significance of another experiment conducted under the same conditions, we must continue our geometrical analysis of the plane mirror's behavior.

If there are two point sources, the waves emanating from each are independently deviated according to the same rule. Then the observer's eyes receive two trains of waves. But it is as if, in place of the mirror and the two sources, there were two other sources situated symmetrically to these with respect to the mirror, the mirror itself being absent.

Instead of two, there may be three, four, . . . *n* sources, grouped so as to constitute lines, surfaces, and objects. Then the mirror produces the following effect. The reflected waves are propagated as if they emanated from an equal number of sources located behind the mirror in positions symmetrical to those of the actual sources with respect to the plane of the mirror. The observer receiving these waves is, therefore, induced to create effigies symmetrical to those that he would have created had he received the waves directly from the actual sources. The placement of the symmetrical effigies thus created is subject to all the rules and uncertainties previously discussed in regard to direct vision. As in that case, the placement turns out to be definite and accurate if the sources are a few meters away from the eyes. When they are farther away, there may be notable discrepancies.

**130.** The conditions of the experiment described in § 128 were modified by replacing the point source with an extended object familiar to the observer—his own face. Previously he had received the waves emitted by the point source and reflected back by the mirror 18 m away. Now, by a small sidewise motion, he took a position with his face on the perpendicular drawn to the center of the mirror, while an electric lamp illuminated his face.

The observer found himself in the awkward situation of being obliged to locate an effigy symmetrical to his own face. The situation was awkward because his mind knew that 18 m away there was a thick opaque wall with no holes in it. Had he located the effigy on the mirror 18 m away, as he had done with the point source, he would have had to attribute to it half of the real linear dimensions, which he knew very well, in order to make them correspond to the angular magnitude of the waves received by his eyes. On the other hand, if he wanted to give the proper dimensions to the effigy, he would have had to place it 36 m away; that is, he would have had to imagine the wall pierced by a hole 16 cm in diameter, outside which there was only his own illuminated head surrounded by darkness, a collection of absurdities too hard to swallow.

The experiment showed that all the observers chose a sort of compromise, a solution midway between the two extremes indicated above. The face effigy was placed about 10 m behind the mirror and consequently was a little smaller than it should have been.

Another noteworthy result was that if the same observer shifted slightly sidewise in the direction opposite to his previous motion, so that he again received the waves emitted by the point source, he located the star effigy on the mirror once more.

Conditions of this kind are, however, extremely rare in practice. Plane mirrors are used rather commonly, but at very short distances from the objects and the observers. The latter, therefore, by effective employment of binocular vision and temporal parallax, almost always locate the effigies with good accuracy behind the mirror in positions symmetrical, with respect to the plane of the mirror, to those of the material sources of the waves.

**131.** A new fact has presented itself to us. In Chapter III the distinction between the source of waves and the effigy may have seemed artificial, and some readers may have thought they would get along without it. Although cases could be cited in which the effigy was certainly located in a position different from that of the source of waves, this was always a result of the difficulty of determining the distance of the source, a difficulty due to the inadequacy of the observer's telemetric means. If he could measure the distance better, the placement would be more accurate. Essentially this was a question of an experimental error.

But now it is not. Given all the desired perfection of the telemetric means, the effigy is still located at a point different from the one in which the source of waves is. For, as was said in § 127, when everything is done perfectly, the effigy is placed at the point symmetrical to the source with respect to the mirror. Then if there are telemetric inadequacies and complications, this rule too is no longer valid. Therefore the effigy is placed in exact coincidence with the source of waves only when this happens to be on the mirror, a case of no interest.

**132.** In the standard language of optics it is customary to say that an observer *sees an object* when his eyes receive waves directly from the sources. It is also said that the aggregate of the centers of the waves reflected by a plane mirror constitutes the image of the object. It is therefore said that when an observer receives these reflected waves, he *sees the image* of the object given by the mirror. At first blush this phraseology appears innocent and so routine that it has been accepted throughout the world for centuries. Yet in reality it implicitly contains dangerous ideas and hypotheses. I shall expose its weaknesses in § 256.

But now I invite the reader to recall, as was said in § 92 about direct vision, that what the observer sees is an effigy created by himself. This may coincide with the source and may also be distinct from it. Consequently, as was then remarked, in all propriety we should avoid saying that the observer sees the object. For to say so amounts to assuming offhand that the effigy and the object coincide, and this may not be true.

By the same token, to say that an observer sees the image of an object given by a mirror may lead the reader to think that the mirror gives an image even before the observer intervenes. This is absolutely not true. A mirror deviates incident waves, and that is all it does. Moreover, the reader may be induced to believe that when an observer looks at a mirror, he sees an image behind it. This too is false. It is the observer who places the effigy behind the mirror. Before he intervenes, so far as optics is concerned, there is nothing behind it.

Accordingly there is no difference between what an observer does when he *sees an object* and what he does when he *sees an image*, in the usual sense of that term. A difference may perhaps be discovered by resorting to verification by touch. This *may* confirm the existence of sources of waves in the former case, but in the latter case there is never any confirmation. Hence, lest the logic of our argument be impaired, I urge the reader systematically to shun the italicized expressions, at least until we have defined them in §§ 257–258 with the greatest care and eliminated all possibility of misunderstanding by those who use them, as many people do, without knowing the fundamentals of the subject.

**133.** To conclude this cursory survey of the optical function of a plane mirror, its effect from the point of view of energy need hardly be mentioned. The reflected waves are not only deviated by the mirror but also weakened. Their amplitude is smaller than that of the incident waves. For, as was pointed out in § 76, any material body reflects only a fraction of the radiant flux which it receives, $\rho$ being always less than 100 per cent and in general a function of $\lambda$. As a result, the effigy created by an observer who receives the reflected waves must have a brightness inferior to what he would have created had he received the waves directly from the source. Also the hue and saturation may be different. But all this is a palpable observation, on which there is no occasion to dwell.

**134.** Let us now go on to analyze what happens when waves emitted by a material source are reflected by a curved mirror.

As our first case, consider a perfectly spherical concave mirror in front of which there is a point source of visual waves. Let the mirror's radius of curvature be $r$ (Fig. 42). As is well known (§ 311), an incident spherical wave of radius $x$ ($x$ being the distance from the source $S$ to the mirror) is reflected as a spherical wave of radius $x'$, defined by the relation

$$\frac{1}{x} + \frac{1}{x'} = \frac{2}{r} \tag{1}$$

or

$$\xi + \xi' = 2\rho$$

where $\xi$ and $\xi'$ are the reciprocals of $x$ and $x'$ respectively, and $\rho$ is the reciprocal of $r$ or the curvature of the mirror.

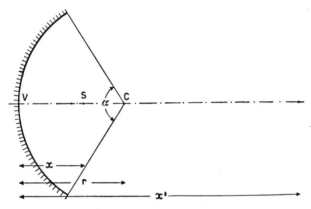

FIG. 42. Geometrical elements of a concave spherical mirror

To be more accurate, we should say that in general the reflected wave is not spherical, although it has symmetry of revolution around the *axis of the mirror*, that is, around the straight line joining the source $S$ with $C$, the mirror's center of curvature. But the deviations from a sphere, which are known as *spherical aberration*, are neglected in this analysis because they are of slight importance; or rather, they become of slighter importance as the mirror's *angular aperture* $\alpha$ diminishes, $\alpha$ being the angle subtended by the diameter of the mirror's rim at $C$, the center of curvature. For the sake of simplicity, we shall assume that the rim of the mirror is a circle with its center $V$, called the *vertex*, on the axis.

**135.** Discussion of the phenomena connected with this type of mirror is highly interesting and productive of arguments that support the thesis I am propounding. Hence I shall pursue this subject at some length. The situations available for study are numerous and deserve to be examined one by one.

To begin with, assume that the source $S$ coincides with the mirror's focus $F$, the point midway between the vertex $V$ and $C$, the center of curvature (Fig. 43). Substituting $x = \frac{r}{2}$ in equation (1), we have $x' = \infty$; that is, the waves reflected by the mirror are plane. They may reach an observer's eye or both his eyes. Suppose the axes of these eyes are perpendicular to the waves. Then in both

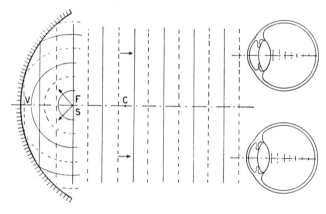

Fig. 43. Binocular vision of a point source located at the focus of a concave spherical mirror

foveas cones are stimulated and send impulses along the optic nerves. These impulses are analyzed by the mind with the aim of creating the corresponding effigy.

Ignoring for the present questions relating to the frequency and modulation of the impulses, that is, those having to do with the brightness and color of the effigy, let us scrutinize those pertaining to its form and position. The optical information reaching the mind may be summarized as follows: the source of waves has minimal dimensions; there is no accommodation or convergence; and the direction from which the waves arrive is the same as that of the axes of the eyes. If the mind proceeded to the end of its task on the basis of the information received by the optical route, it would have to create a point effigy or star in the direction of the axes of the eyes at a very great distance behind the mirror, exactly as an observer does when he looks at a star in the sky.

But this result is seldom found. When it is, it must be regarded as an extremely exceptional case. The reason is obvious. The information of an optical nature is superseded by other environmental reports, which the mind takes good care not to ignore.

**136.** To examine all the cases that may be met is impossible in practice because they are of infinite variety. We shall analyze only a few of them in order to acquaint the reader with the nature of the circumstances affecting the outcome.

Suppose the source is so perfectly screened and covered that the observer cannot receive direct waves from it or its container. Consequently he has not the least idea of its existence and location. Assume also that the mirror is highly polished and free from optical defects, that its rim is invisible, and that the background is absolutely dark. These are the simplest conditions. In the entire setup the only visual waves are those that go from the source to the mirror and, after being reflected there, reach the eyes of the observer. He should be brought into the laboratory when it is completely dark to prevent his having any conception of its layout.

Despite all these precautions and others that the experimenter may devise, nobody can stop the observer from thinking that he has entered a dark place, whose dimensions he imagines to be like those of an ordinary room, since he has no means of estimating them in the present instance because everything is dark and black. Therefore, when the waves reflected by the mirror reach his eyes, he cannot persuade himself to locate the star effigy at a very great distance, for he is aware that he does not have anything like the celestial sphere in front of him. He puts the star in the direction defined by the structure of the waves, but at a distance of just a few meters. It is impossible to say any more. The placement is performed by each observer according to his mind's mode of thought at the moment of the observation.

Another action often occurs in the course of this experiment. The observer moves his head in the direction perpendicular to the axis of the mirror, as he usually does in resorting to temporal parallax. If the effigy were situated at a very great distance, obviously it should not move no matter how much the observer's head is displaced. But it is generally seen to move just like the head, because it has been located too near.

If the experiment is performed later without these precautions, so that the observer receives waves directly from the source too and in addition is permitted to see the mirror, the room and all the details, the final result is even more uncertain and varied according to the circumstances and, among other things, the curvature of the mirror and its distance from the observer.

The most common result is like the previous one. Nobody places the star at a very great distance, as if the mirror were the mouth of an immensely long tube protruding through the walls of the laboratory and everything outside. This notion is so absurd that no mind entertains it. Consequently all the placements are made at nearby distances, even if these are contrary to the information supplied by the organs of sight.

**137.** Let us now move the source closer to the mirror by any interval whatever, say $r/6$. Then $x = r/3$, and therefore

$$\frac{3}{r} + \frac{1}{x'} = \frac{2}{r}; \qquad x' = -r$$

Hence the waves reflected by the mirror are divergent with their center $I$ behind the mirror at a distance from it equal to its radius of curvature (Fig. 44).

If an observer's eyes receive these divergent waves, he must accommodate a little in order to reduce the number of affected cones to a minimum, and he must also make the axes of his eyes converge to fuse the two effigies into one. Two small groups of cones at the center of the two foveas $F_d$ and $F_s$ are thereby stimulated.

The information now reaching the mind resembles the report arising from the case illustrated in Fig. 43. But there is a slight difference, because accommodation and convergence have been invoked. These adjustments concur in locating the center of the waves at a given distance behind the mirror on its axis. The placement may even be made on this basis, but that is only one possibility. It is generally made on the basis of the environmental factors that operated in the previous situations, and it is made at the same distance from the mirror as before.

Direct proof of this is available at once. While the observer is looking at the mirror and receiving the waves reflected by it, move the source $S$ along the axis until it returns to $F$, where it was in Fig. 43. The curvature of the reflected waves steadily diminishes until it disappears; in other words, their center $I$ behind the mirror recedes to infinity. This is not noticed by the ordinary observer. For him, the effigy remains stationary behind the mirror, where he had placed it at the beginning of the observation.

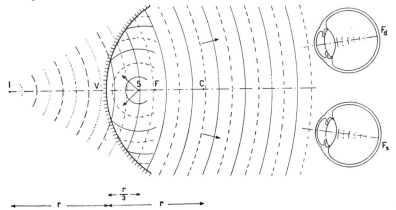

FIG. 44. Binocular vision of a point source located between the focus and the vertex of a concave spherical mirror

What I have just reported is the outcome of the usual observations. It is not unknown that things should go differently and even unexpectedly. I wish to reiterate that the placement is made by the mind of the observer. Any mind, as a result of strange and special influences, may come to the most absurd conclusions in the world, even without encroaching on the realm of the abnormal.

When the source is brought very close to the mirror, the center of the reflected waves comes so near to the source that the placement is of course influenced thereby. But this is a case of little interest.

**138.** Much more interesting, on the other hand, is what happens when the source is moved away from the focus $F$ toward $C$, the mirror's center of curvature. Since the possible cases become even more numerous and varied, the conditions of the experiment must be described in greater detail. For instance, let the radius of curvature $r = 1$ m, while the observer is stationed with both eyes at 2.5 m from the mirror (Fig. 45). Move the source $S$ 4 cm from the focus $F$ in the direction of $C$, so that $S$ is 54 cm from the vertex $V$. In these circumstances the spherical waves emitted by the source are reflected by the mirror as spherical waves with their center at a point $I$ 6.75 m away from the mirror, or 4.25 m behind the eyes of the observer.

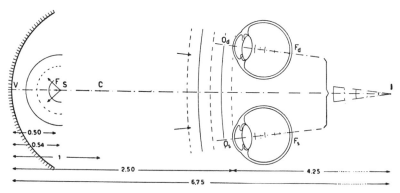

FIG. 45. Binocular vision of a point source located between the focus and the center of a concave spherical mirror

He finds himself in a novel and disconcerting situation. As was indicated in § 135, the eyes keep the crystalline lens passive when they receive plane waves. But if they receive *convex* waves, they make it increase its strength (§ 97). This is greater, the more convex the waves are or the nearer the source is (§ 98). This relation provides the basis for measuring the accommodation of the crystalline lens in diopters, the essential variable being the vergence of the source with respect to the eyes (§ 96). Obviously it makes no sense to talk about negative accommodation (that is, accommodation less than zero) since the normal eye is constructed for direct vision.

When *concave* waves impinge upon a normal eye, there is no internal mechanism to compensate for their curvature, as happens within certain limits for convex waves. Hence the convergent wave within the eye has its center in front of the retina. Therefore the number of retinal cones affected is greater than the minimum, with the excess becoming larger as the curvature of the wave incident upon the eye increases. The effect should be the creation of an effigy bigger than a point, as in the case where accommodation is inadequate (§ 104) or where it is excessive as a result of certain pathological conditions.

But variations of $1/4\ D$ in the curvature of a wave are not perceptible, as was pointed out in § 99. In the example under consideration now, the curvature of the wave incident upon the observer's eyes is slightly less than $1/4\ D$. Consequently the form of the effigy created by the observer's mind will not differ from what the mind fashioned when the source $S$ was at the focus $F$.

**139.** Even stranger is what happens to convergence in this instance. To get the group of stimulated cones into corresponding points exactly at the center of the two foveas $F_d$ and $F_s$, as is required for fusion of the two effigies connected with the two eyes, their axes must *diverge* in order to pass through $I$, the center of the reflected wave. This is a movement to which normal eyes are not adapted, because it need never be executed in direct vision. Nevertheless the muscles are not so rigid that they cannot perform an unusual movement, at any rate for a brief period of time, a sensation of fatigue being felt very soon. In the present circumstances, in which the angle to be formed by the axes is barely $48'$ (for an interpupillary distance of 60 mm) almost all observers are able to have their eyes form this angle, without any great satisfaction, but also without any undue strain.

The mind, however, is put in a piteous predicament. The waves of course enter the eyes from in front, but convergence (which now is *divergence*) reports that the source is located in back of the head. This announcement, being devoid of significance from the optical point of view since no observer is able to locate effigies behind his own head, is ignored. The star effigy is placed out front in the direction of the line bisecting the angle formed by the axes of the eyes, at the distance deemed most reasonable by the mind.

If the source $S$ is moved to a position between $F$, the focus of the mirror, and $C$, its center of curvature, after having been on the other side of $F$ as well as in $F$, the effigy still continues to be assigned to the same spot, even though the waves have changed from convex to plane and then to concave; in short, the mind pays no attention to the information based on accommodation or convergence.

**140.** Let us now move the source $S$ farther away from the focus toward the center $C$, say, to a point 60 cm from the mirror. The reflected waves are still concave, with their center at a point $I$ 3 m from the mirror or just 50 cm behind the eyes. Hence the waves have a vergence of $2D$ with respect to the eyes.

The discussion proceeds as in § 138, but the conclusions are quite different. The stimulated retinal surface, now distinctly larger than the minimum, contains several hundred cones (and also rods). The divergence necessary for fusion amounts to almost 7° (412', for 60 mm of interpupillary distance) and only rarely can anyone be found who is capable of attaining it.

The mind, accordingly, is seriously embarrassed. It must create two effigies that it cannot fuse. It has to make them rather large in size (and circular, if the pupil is circular; but as a result of irradiation, which is always present, they assume the appearance of enormous radiated stars) and has no optical reason for locating them at any particular distance.

Under conditions of such complexity, observers generally rebel and shut their eyes. If they are very sensitive, they even experience feelings of nausea and dry vomiting. If they are trained in this kind of observation, they usually close one eye and observe with the other; that is, they create a single effigy of appropriate magnitude and place it before the open eye at an entirely arbitrary distance. If the observer is brought into this situation by a continuous transition out of the preceding situations, he sees the effigy expand while remaining at the same distance behind the mirror.

**141.** Let us now move the source 625 mm away from the mirror (Fig. 46). The reflected waves are even more concave than before, and have their center 2.5 m away from the mirror, exactly where the observer's eyes are.

In this position the wave, being reduced to a very small centric, cannot affect both eyes at the same time. Since the centric is on the axis of the mirror, the eye that the observer wishes to use must also be placed on the axis. Thus the convergent wave enters the eye without undergoing any appreciable deformation, and impinges upon a retinal zone corresponding to the surface of the mirror. The boundary of this zone and the rim of the mirror are obviously homothetic.

The information reaching the mind is such that it is led to create an effigy consisting of a luminous disk (for we have assumed that the rim of the mirror is circular). The disk's angular magnitude is defined by the dimensions of the affected retinal zone. Induced to see a figure having the form of a disk, the mind tries to make the

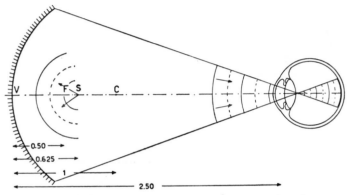

Fig. 46. Point source seen in a concave spherical mirror when the center of the reflected waves falls within the eye

disk's edge brighter by means of accommodation. Hence the eye sets its accommodation for the mirror, in this case employing $0.4D$ of accommodation. In the end the effigy created is a luminous disk (at the very most there will be at its center a dark spot due to the presence of the container enclosing the source of waves) located where the observer believes the mirror to be. In ordinary language he is said to "see the mirror all lit up."

**142.** Let us continue to move the source $S$ until it is 67 cm away from the mirror (Fig. 47). The reflected wave, which is still concave, now has its center at a point $I$ 2 m from the mirror or 50 cm in front of the observer's eyes. These must therefore exert $2D$ of accommodation to reduce the number of stimulated retinal elements to a minimum, and must also converge almost 7° to avoid diplopia.

Hence the information reaching the mind would impel it to conclude that there is a point source of waves 50 cm from the eyes, and accordingly to create an effigy in the form of a star exactly 50 cm in front of the nose. All this would be fine if the mind did not reject the idea that in front of the nose there could be a luminous point suspended motionless in the air. To see it in this way, that is, to locate the star effigy in this manner, the mind must first be convinced that such a thing can exist. The observer may be tempted to

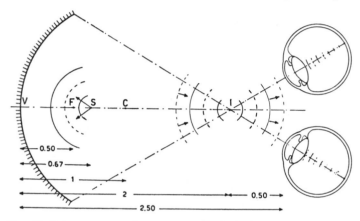

Fig. 47. Point source seen in a concave spherical mirror when the center of the reflected waves falls in front of the eyes

stretch his arm out to ascertain whether his hand touches the source of waves.

Usually none of this happens. For it is an ingrained habit in looking at a mirror to see figures behind the reflecting surface since plane mirrors, consisting of the calm surfaces of liquids if of nothing else, are frequently encountered in everyday life. The ordinary observer, when put before a concave mirror, is dominated by the preconception that the effigy must be located behind the reflecting surface. This preconception is not overcome even by the clear and definite information provided by accommodation and convergence. As a result the star effigy is placed behind the mirror.

Following this conclusion, the eyes tend to direct their axes toward that effigy 3 m or more away, and consequently diplopia results. The eyes are then induced to converge on a nearer point to achieve fusion. A contradictory and extremely unpleasant situation ensues. The observer shuts his eyes. Sometimes he has feelings of nausea and vomiting. Generally he ends up either by looking with only one eye (placing the effigy behind the mirror, now that convergence is no longer a disturbing factor and accommodation is discounted) or by turning the gaze of both eyes in another direction.

The effects may also be quite different. For example, near $I$ in Fig. 47 put any material object, say a stick with one end beneath $I$, so as to give the impression that it supports the source of waves, or a diaphragm perforated so that the wave passes through the central portion of the aperture. These arrangements yield information that is completely at variance with the previous reports and entirely plausible to the mind. It no longer has any difficulty in placing the star effigy at the point indicated by accommodation and convergence. This trick is necessary for the ordinary observer, but superfluous for an experienced person who is convinced that he should place the star effigy in the air.

**143.** If we continue to move the source $S$ farther away from the mirror, the situations that arise do not differ substantially from the case analyzed in § 142. The ordinary person generally finds the observational conditions less uncomfortable than those described there. For the center of the reflected waves approaches the mirror and recedes from the eyes, thus permitting a reduced effort of accommodation and a smaller convergence. Hence the observer's mind encounters weaker resistance to satisfying its desire to place the star effigy behind the mirror.

For example, if the source $S$ were removed to a very great distance, the waves transmitted to the mirror would be sensibly plane and the reflected waves would have their center at the focus $F$. The observer's eyes, still 2.5 m away from the mirror, would need only $0.5D$ of accommodation and only $100'$ of convergence. In these circumstances the ordinary person usually places the star effigy on the mirror or a little behind it.

It would be instructive also to discuss the application of temporal parallax by the observer under the various conditions that we have investigated. But I believe that we have devoted enough time to vision of a point source with a concave mirror as intermediary.

Let us go on then to examine what happens when the source of waves consists of an object of appreciable dimensions.

**144.** For reasons that will become clear later, it is convenient to consider linear objects perpendicular to the axis of the mirror. Or rather, we shall assume that we are talking about arrows (as indeed they will be drawn in the Figures) with one end on the axis of the mirror. Any such object may be regarded as an assemblage of points, each of which emits waves on its own account. Every train of waves is reflected by the mirror and returns from it as another

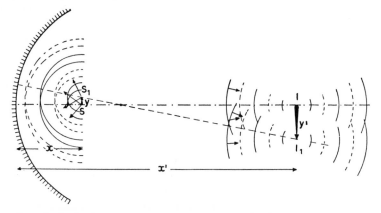

FIG. 48. Reflection by a concave spherical mirror of waves emitted by various elements of an extended object

train of waves that have somewhat complex forms. But in this part of our investigation we should side-step those complications by treating the reflected waves also as spherical. The difference between such a treatment and the actual facts diminishes as the angular aperture of the mirror becomes smaller and the object under consideration becomes shorter.

Simplifying by hypotheses of this kind, we arrive at the equation (§ 312)

$$\frac{y}{x} = \frac{y'}{x'}$$

where $y$ is the height of the object or the distance between its end points $S$ and $S_1$, while $y'$ is the analogous distance between the centers $I$ and $I_1$ of the waves reflected by the mirror and emanating from the sources at the two extremities, respectively (Fig. 48).

The fact that the source is not a point but an object, often of known dimensions, supplies the mind with new criteria that play no small part in locating the effigies. Exactly as in direct vision (§ 121), the effigies of known objects are placed much more rapidly and accurately and at far greater distances than are the effigies of unfamiliar and unknown objects; the effigies of known objects, moreover, provide highly important standards of reference for the placement, by suitable study of the parallaxes, of effigies otherwise hard to identify. As we noticed in the entire discussion of the point source, in vision by means of plane and curved mirrors the localization of the effigies is a very complex problem. Introducing the dimensions of the object is, therefore, not without interest.

**145.** Let us examine, for example, the case in which the object lies in the focal plane (Fig. 49). Let $y$ be the length of the object.

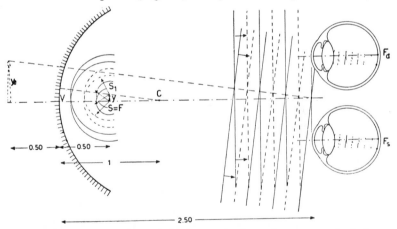

Fig. 49. Vision of an extended source at the focus of a concave spherical mirror

For our present purpose we need consider only its end points as sources of waves; the points in between obviously behave in an intermediate manner.

Let the source $S$ coincide with the focus $F$ on the axis of the mirror. The waves that the source emits, as we saw in § 135, leave the mirror as plane waves perpendicular to the axis. When they encounter the eyes of the observer (he is still 2.5 m away from the mirror, whose radius of curvature is 1 m) they enter and are made convergent with their centers at the two foveas $F_d$ and $F_s$. The axes of both eyes are of course parallel to each other and to the axis of the mirror.

In precisely the same way the waves propagated by the point $S_1$ toward the mirror are reflected back in the form of plane waves. To discover the disposition of these waves we need only consider that the straight line $S_1C$ may be regarded as an axis of the mirror with just as much right as the straight line $SC$. This latter line is called the *principal* axis of the mirror, simply because it passes through the point $V$, the center of the mirror's rim. After the waves emitted by $S_1$ are reflected, they must be propagated in the direction of the straight line $S_1C$. Hence when they reach the observer's eyes, they are made convergent with their centers at two retinal points corresponding to that direction. Thus the angular magnitude of the effigy created at the end of the visual process is constant, independent of the observer's distance from the mirror, and exactly equal to the angle $SCS_1$ subtended by the object at the mirror's center of curvature.

**146.** The information now furnished to the mind by accommodation and convergence is that the source of waves is extended, linear, and very far off. There may also be reports concerning the brightness and hue of the various elements of the effigy. We shall not take account of these two qualities, *even though they are not always negligible* for the purpose of placing the effigy.

This optical information is submerged, however, beneath other environmental or experimental considerations that lead the observer to locate the effigy rather near. But where? That obviously depends on the observer's frame of mind at the moment of the experiment. Hence a universal answer cannot be given. Fairly commonly the effigy is put in back of the mirror at a distance equal to the distance of the object from the front of the mirror, in the present instance, 50 cm. This placement is suggested to the observer by his long experience with plane mirrors, as was indicated in § 142.

Then the effigy is localized 3 m from the observer, and its angular magnitude is the same as that subtended by the object at the center $C$. Therefore $y_e$, the length of the effigy, will be six times the length of the object, this being the ratio between the distance (300 cm) from the effigy to the observer and the distance $SC$ (50 cm) from the object to the mirror's center of curvature. But there is no assurance that the observer sees the effigy so many times larger than the object, since in that space before the eyes angular magnitude is considerably underestimated. However, if the observer is in a position to receive at the same time the waves coming directly from the object, he must create the two effigies simultaneously, and then they are usually constructed in proportion to their respective angular magnitudes.

**147.** Hence we are now confronted by this new phenomenon, *magnification*, as it is called. It is a very interesting topic that we must treat at length, both here and in connection with other optical systems too. Many have tried to define it, but heretofore they have had to fall back on conventions of a geometrical kind.

Anyone who wants to know how much magnification a certain optical system provides is asking, essentially, how many times the figure that he sees with the help of that system is bigger (or smaller) than what he would see by looking directly at the object. Merely to remove a possible cause of confusion, I point out that this ratio may prevail between two corresponding lengths or between two corresponding surfaces. The former ratio is known as *linear magnification* and the latter as *areal magnification*. Simple geometrical arguments lead at once to the conclusion that the latter is numerically equal to the square of the former. It is, therefore, unnecessary to consider them both. Only the linear magnification is used at present, according to the established practice that we shall follow hereafter without the need of so specifying on each occasion.

Returning now to the problem of defining magnification, I wish to emphasize that the previous question cannot have a precise answer of a mathematical or physical nature. The figure that anyone sees when he looks at an object is an effigy created by the observer. The figure that he sees when he looks in an optical system is another effigy created by his own mind. To inquire how many times the latter effigy exceeds the former in length would have meaning if the human mind acted uniformly, if the mind of any observer always functioned in the same manner, and if this manner were connected with the physical characteristics of the objects without any subjective intervention.

All three of these suppositions have to be rejected. Then there is no sense in asking how many times the figure seen by means of an optical system *must* exceed that seen directly. If anyone desires this information in a particular case, there is only one way to get it—interrogate the observer and trust his answer. *For optical magnification is an essentially psychological operation.*

Nevertheless some patterns of behavior have a high probability of being found in the majority of observers. In §§ 148–151 I shall deal with a few of these patterns under various circumstances. If anyone trying an experiment obtains different results, he should not be surprised. The variation simply means that at that moment some particular piece of information induced his mind to make a decision different from the usual.

**148.** Thus, in the example considered in §§ 145–146, if the observer receives at (or about) the same time the waves emitted directly by the object 2 m from his eyes as well as the waves reflected by the mirror, he creates two effigies. Of these, the former may be placed with precision, especially if the object is well known. On the basis of the optical information, the latter effigy should be located at an enormous distance and accordingly have gigantic proportions. In that case the magnification (in other words, the ratio between corresponding dimensions of the two effigies, as was explained in § 147) would be tremendous or, to use the conventional expression, infinite. But this never happens. Frequently the effigy is stationed behind the mirror at a distance approximately equal to the object's distance (50 cm) from the mirror. Hence the ratio is 6 (as was shown by the calculation in § 146). A direct comparison between the two dimensions being possible, this ratio may be estimated fairly accurately.

But if only the reflected waves reach the eyes, and the object is familiar, the mind finds itself in one of its customary embarrassing predicaments. Both the optical and geometrical information would induce it to construct an effigy magnified much more than six times. Then the conclusion depends on the observer's mode of thought. In many instances he works out a sort of compromise. The effigy is made bigger than required by the memory, but smaller than indicated by the optical information, smaller by about half. Accordingly there is no reason to wonder, if the observer who is put in this situation usually reports a magnification considerably below what would be expected.

There are effects, however, that almost all observers notice. Thus it readily follows from our previous analysis that if the observer approaches the mirror, the effigy seen behind it diminishes proportionally. For the effigy's angular magnitude is constant, being equal to the angular magnitude subtended by the object at the center of curvature. The position of the effigy behind the mirror is likewise fixed. Therefore the effigy's dimensions must vary in proportion to the distance between the observer and the effigy; that is, they must decrease if the observer approaches the mirror, and increase if he recedes. Even though these observations are made in the zone where the mind attributes no great importance to the angular magnitude, nevertheless by and large this effect is actually observed.

**149.** For the sake of brevity I shall refrain from analyzing what happens when the object is displaced from the focus $F$ toward the mirror, because not much is added thereby to what was said about the case considered in §§ 145–148.

Let us instead move the object from the focus $F$ toward the center $C$ until it is 54 cm from the mirror (Fig. 50). The only essential difference between this situation and the one illustrated in Fig. 49 is that now the waves reach the observer's eyes slightly concave. To avoid diplopia, the axes of the eyes must diverge a bit, as was mentioned in discussing a similarly located point source (§ 139). The mind is even more embarrassed than before, particularly if the eyes receive, simultaneously with the waves reflected by the mirror, those coming directly from the object about 2 m away. The $0.5D$ of accommodation necessary for direct vision of the object is inconsequential and may be disregarded. But to avoid the diplopia resulting from the waves reflected by the mirror, the eyes must diverge almost 48', as in the case of the point source (§ 139). On the other hand, to avoid the diplopia caused by the direct waves, the eyes must converge as much as 105′. In other words, to pass from one setting to the other, the convergence must change by $2\frac{1}{2}°$. Under these conditions the mind is truly embarrassed in placing the effigies.

Yet the conclusion is simple. Ignoring the optical information, the mind locates the effigy a little behind the mirror, just as when the object passed through the focus (§ 146). The magnification is a trifle larger. But the difference is appreciable only if the object is

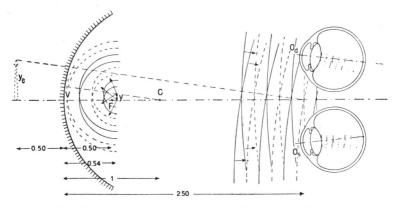

Fig. 50. Vision of an extended source placed between the focus and the center of a concave spherical mirror, with the centers of the reflected waves falling behind the eyes

moved rather rapidly while the observer is receiving the waves reflected by the mirror.

Under these conditions too, if he moves backward while looking at the mirror, the effigy created is enlarged, but somewhat more quickly than in the case treated in § 148.

**150.** Now let us move the object farther away from the focus until it is 62.5 cm from the mirror (Fig. 51), the same distance as in Fig. 46. The waves emitted by any point of the object, after being reflected by the mirror, become concave with their center 2.5 m from the mirror or exactly where the observer's eyes are.

If he is still, his right eye remains inactive because it does not receive any reflected waves. But his left eye is entered by some waves, namely, by those whose centers coincide with the pupil.

Each such wave stimulates the retinal elements contained in a disk, just as when the source was a point (Fig. 46). The effigy created will, therefore, be a luminous disk as big as the mirror and located on the mirror (unless the observer errs or thinks otherwise). There is no longer any trace of the object's form.

What is seen when the object is moved from the focus $F$ toward its position in Fig. 51 is peculiar. As was indicated in § 149, when the object passed from 50 to 54 cm away from the mirror, the magnification increased; in other words, the effigy stationed behind the mirror acquired bigger dimensions and might be said to occupy a

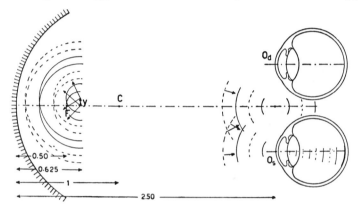

Fig. 51. Extended source seen by means of a concave spherical mirror when the centers of the reflected waves fall within one eye

larger portion of the mirror. As we continue to move the object farther away from the focus, the effigy enlarges until, when the object reaches a distance of 62.5 cm, as in Fig. 51, the entire mirror seems to be filled by it.

The magnification may be said to have become enormous. During this process the effigy gains in size but loses in brightness. This loss is due to the fact that the reflected waves reach the eyes concave with a progressively greater curvature, on account of which the waves refracted within the eye have their centers farther and farther away from the retina.

**151.** As we continue to move the object farther away from the mirror, the effigy that seemed to take up the whole mirror starts to shrink and shows another peculiarity by turning upside down.

To account for what is seen in these circumstances, let us halt the object 67 cm from the mirror under conditions like those in Fig. 47. The waves emitted by each point of the object, after being reflected, become concave with their centers 2 m from the mirror or 0.5 m from the observer's eyes (Fig. 52). Every center $I_1$ of a reflected wave is collinear with the corresponding emitting point $S_1$ and also with $C$, the mirror's center of curvature. Consequently, if the point $S_1$ is above the axis $VC$, the point $I_1$ must be below it.

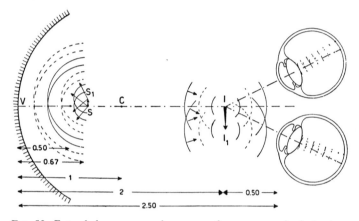

Fig. 52. Extended source seen by means of a concave spherical mirror when the centers of the reflected waves fall in front of the eyes

The eyes now receive divergent spherical waves coming from all the points on the line between $I$ and $I_1$, as if there were a real object between these extremities. With $2D$ of accommodation and nearly $7°$ of convergence, the stimulated retinal zones have minimal dimensions, and diplopia is avoided. All the optical information would lead the mind to station an effigy in the form of a line, situated like $II_1$, 50 cm in front of the eyes.

But this generally does not happen. Such an effigy up in the air is contrary to common sense, and no one (but an initiate) can conceive it as possible. The effigy must be snspended from, or supported by, something. I say nothing about the complications arising when the object $SS_1$ is a familiar thing. If it is a glass full of water, for instance, there is no observer capable of placing at $II_1$ an effigy consisting of an inverted glass wherein the water remains clinging to the bottom that is on top.

Usually the observer locates the effigy on, or a little behind, the mirror. The effigy of course comes out much bigger, considering the greater distance to which it is removed while its angular magnitude remains constant. The resulting situation is so strange and absurd that the observer's mind no longer knows what information to trust.

Here again a most unpleasant conflict breaks out between the anticipation derived from the memory and the information furnished by the optical machinery. The mind wants to put the effigy behind the mirror. Then the optical axes are brought to converge on this effigy. But such a movement does not agree with the waves actually reaching the eyes, and therefore diplopia results. To avoid it, the optical axes must be made to converge on $I$. But this action does not agree with the anticipation derived from the memory. The observer protests; shuts his eyes, or one of them at any rate; has queasy feelings; and reports nausea and dry vomiting as well as a strong urge to look elsewhere.

**152.** The situation becomes normal at once if some material object is put at or near $I$ to eliminate the absurdity of an effigy sus-

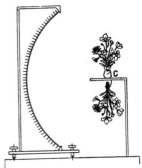

FIG. 53. Experiment with a bouquet of flowers

pended in mid-air. It is then assigned a position very readily on the basis of the optical information.

In this connection the "bouquet of flowers" is a very old experiment (Fig. 53). Below $C$, the mirror's center of curvature, place a bunch of flowers upside down. Light them up on the side facing the mirror, but conceal them by an opaque screen on the side facing the observer. The centers of the reflected waves lie above $C$ in inverted order. To facilitate the placement of the effigy, set a vase at $C$, with no flowers in it, of course. Any observer sees with complete assurance a vase with a bouquet of flowers. The closer he approaches, even to 20 or 30 cm, the more convinced he is that the bouquet is there. Some people stretch out their hands to take the flowers and are surprised to find only the vase. The flowers that they saw with such confidence were effigies created by themselves. Naturally, in order to be able to do so, the observers must direct their eyes toward the course of the waves reflected by the mirror.

**153.** The method with which I have sought to approach the study of vision by means of a concave spherical mirror has now been expounded at some length. Yet only two equations were introduced into the discussion, which was limited to a few special instances. Rather than state any general rule, I intended to offer an example. To anyone who has understood its essence, it may serve as a guide whenever a similar case appears. The absence of a mathematical treatment is nothing but an indication of how subjective and arbitrary this process of vision is. Yet experimental observations and what I have set forth correspond to a surprising extent. The surprise is all the greater because the type of reasoning I used is deemed by many to be almost heretical or at least erroneous. But in § 192 I shall demonstrate how erroneous is that which is presented everywhere today as the most certain truth.

**154.** Let us now go on to a brief consideration of vision by means of a convex mirror.

As is well known, when the waves emitted by a point source are reflected by a convex mirror, they become more divergent. Their center of curvature is, therefore, behind the mirror at a point whose distance $x'$ from the mirror is given by the equation

$$\frac{1}{x} + \frac{1}{x'} = \frac{2}{r}$$

where $x$ is the distance from the source to the mirror ($x$ being taken as positive, because it is in front of the mirror) and $r$ is the mirror's radius of curvature (taken to be negative). Hence $x'$ is always negative and numerically less than $r/2$.

Accordingly, wherever the observer may be, his eyes receive waves that inform his mind that there is a center of propagation at a relatively short distance behind the mirror. This report is never contradicted by information derived from the optical mechanism or the memory, or by the reasoning of common sense. Consequently the effigy will always be located behind the mirror approximately where required by the optical information.

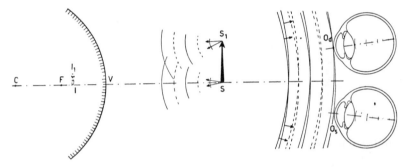

FIG. 54. Extended source seen by means of a convex spherical mirror

This situation is made clear in Fig. 54. The waves emitted by $S$ are reflected by the mirror as if they were emitted by a point $I$ situated between $V$ and $F$; that is, $S$, $V$, $I$, $F$, and $C$ are necessarily collinear. In like manner the waves emitted by $S_1$ are reflected as if they were emitted by a point $I_1$ on the straight line joining $C$ with $S_1$. The distance $II_1$ is always shorter than $SS_1$.

On the other hand, the angular magnitude that the eyes $O_d$ and $O_s$ get from these reflected waves is obviously smaller than the angular magnitude that the eyes obtain from the waves transmitted directly by the object $SS_1$, if these direct waves are received by the eyes. It follows at once that the effigy will be placed behind the mirror, oriented like the object and made smaller than it.

**155.** It is not worth while to linger long over this case. Nor shall we take up mirrors of cylindrical, conical, or irregular design because the time has come to turn to the analysis of vision through a refracting surface. The outstanding example, which has been observed and studied for thousands of years, is that of bodies immersed in water and seen through its unobstructed and undisturbed surface. This situation is still so common today that it lends itself to easy and interesting experiments. Hence I shall use it as an illustration in the discussion I am about to undertake, because as usual I prefer the examination of a particular instance to a general treatment.

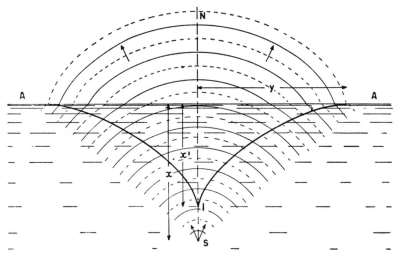

FIG. 55. Refraction of waves passing through a plane surface

When a point source of visual waves lies in the water, its waves are spherical until they reach the surface. Upon emerging into the air, they lose their spherical shape and assume a much more complex structure. Obviously they still form a surface of revolution around $SN$, the vertical line passing through the source $S$ (Fig. 55). But the normals, instead of intersecting at a single point, envelop an evolute surface, called a *caustic*. This is also a surface of revolution around the vertical; its border is tangent to the surface of the water; and it has a cusp on the axis of revolution. If we indicate the distance from the source $S$ to the surface $AA$ by $x$, the radius $y$ of the caustic's border, where the evolute surface is tangent to the refracting surface, is given by

$$y = \frac{x}{\sqrt{n^2 - 1}} = \frac{x}{0.88}$$

$n$ being the refractive index of water or 4/3. The distance $x'$ from the cusp $I$ to the refracting surface is given by

$$x' = \frac{x}{n} = \frac{3x}{4}$$

**156.** When an observer looks at the water, he receives the emerging waves. For the sake of simplicity let us assume his posture to be such that the waves reaching his eyes are symmetrical.

Consider what happens in a single eye. Arriving at the pupil is a small wave segment that varies in accordance with the position of the eye. If this is on the vertical $SN$ and is looking down, the waves impinging upon it are sensibly spherical with their center at the cusp $I$. Inside the eye they are made convergent and, with appropriate accommodation of the crystalline lens, are concentrated at their center on the fovea. The mind thus acquires information about the existence of a center of waves on the axis of the eye at a distance determined by the usual methods.

The cases entirely deficient in reports from which the mind can deduce the distance of the source are so difficult (we are dealing with a point source, be it noted), artificial, and infrequent that there is no need to take them up. Since the source is always joined to some support or to some more or less familiar object, the placement is almost invariably based on acquaintance with the form of the object in complete disregard of the admittedly crude data furnished by the effort of accommodation. This matter will be discussed in greater detail in § 158, when we examine extended sources immersed in water.

Before doing so, however, we should continue analyzing vision of a point source. Instead of using only one eye on $SN$, the perpendicular to the refracting surface, the observer may utilize both eyes in binocular vision, disposing them so that they are symmetrical, or nearly so, with respect to $SN$. Then convergence intervenes to supplement the information conveyed to the mind concerning the distance of the center of the waves incident upon the eyes. But the results do not differ sensibly from those that we mentioned above in connection with monocular vision.

**157.** Of greater interest is what is observed when the direction of vision is oblique. In this case the effects are more perceptible, the larger the angle formed by the axis of the eye with the vertical $SN$ (Fig. 56). For under these conditions the wave segment arriving

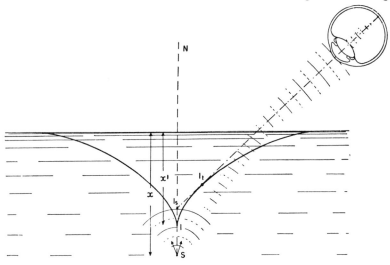

FIG. 56. Point source seen through a plane refracting surface

at the eye is no longer spherical but toric, with its maximum curvature in the vertical plane and minimum in the horizontal. The center of curvature of the vertical section is on the caustic at a point $I_t$, while the other center is on the vertical at a point $I_s$, which is obviously above the cusp.

The eye makes this wave segment convergent, but cannot make it spherical. Despite accommodation, therefore, the group of stimulated retinal cones cannot be the minimum mentioned in previous cases, and the phenomenon known as *astigmatism* arises. Yet its influence on the acuity of vision, a subject with which we shall deal in Chapter V, is so slight under these circumstances that it may be neglected.

I mentioned it solely for the purpose of pointing out that even if accommodation were a potent mechanism of impeccable accuracy, it could furnish only very strange information now. It could no longer indicate that the wave incident upon the eye has a center at a certain distance, but it would have to report that the wave is toric and as such shows two zones of concentration.

The perplexity still remains when binocular vision is employed. If the eyes are at the same distance from the water and symmetrical with respect to a plane passing through the vertical $SN$, they find that they must choose between points of the type $I_t$ and the type $I_s$. On the other hand, if both eyes are not at the same distance from the water, that is, if the observer tilts his head, the situation is even more complex. Were the eyes' telemetric system possessed of absolute precision, it would have to notify the mind that the waves reaching the eyes do not have a center, hence are not emitted by a point source, and therefore a point effigy should not be created.

Not only is the ocular telemetric system not  :rfect, or rather somewhat crude, but its reports are analyzed by a mental apparatus that has astonishing resources and can resolve difficulties with surprising skill. By selecting from the various items of information those that it regards as more coherent and useful, while ignoring the others, it creates a star effigy and places it in the water. But where? It is difficult to say. Most of the time the observers themselves do not know. The only definite statement that can be made is that the effigy is located nearer to the surface of the water, the greater the obliquity of the axis of the eye. As for the rest, there is no need to spend any more time considering a point source. For in the case of extended objects, psychological factors intervene that are more influential than the optical. Hence we should examine some instances of this sort.

**158.** In the usual way, consider an extended object as the aggregate of the point sources composing it. Each of these on its own account emits waves, with regard to which what was said in §§ 156–157 must be repeated.

An observer who receives waves from all these points in both his eyes finds himself in countless different situations at the same time. Suppose, for example, that he turns his face down and looks vertically at the water. Then each eye receives not only sensibly spherical waves from the points of the object that are located on the respective verticals but also toric waves from the adjacent points. The greater the distance of these points from the verticals, the more astigmatic are the toric waves. Moreover, the inclinations are different for the two eyes, which are, therefore, at different distances from the centers of curvature of the waves.

Hence the information reaching the mind is extremely complex and intricate. Amid this hodgepodge it tries to identify the object with which it is concerned, creates the related effigy, and locates it in the water by relying on judgments drawn entirely from the memory. Accordingly the seeing takes place almost as if refraction were not involved. For instance, bowls full of water about a foot deep are used by millions of people who look into them. Nevertheless, no one has ever found that the figures seen by him (that is, the effigies that he localized in the water) were closer to the surface than were the material objects that he proceeded to grasp with his fingers. Yet the displacement must have been several inches.

On the other hand, surprise is occasionally felt in dealing with water two feet or more in depth. Sometimes the observer "sees an object" in the water, thinks he can pick it up with one hand, rolls up his shirt sleeve as far as he believes necessary to avoid wetting it, and then has to get it wet if he wants to succeed in taking hold of the object, which as a matter of fact is much lower than the effigy localized by him on the basis of the optical data.

A noteworthy case occurs when the depth of a pond is estimated from its shore. Suppose the observer naïvely makes his calculation by looking at the objects lying on the bottom. The waves affecting his eyes necessarily reach him from directions markedly inclined to the vertical. Consequently the effigies rise conspicuously closer to the surface of the water. The observer judges the depth to be barely two feet, goes into the water and feels it come up to his neck, if not higher.

This discussion provides a handy explanation of the ancient observation of an oar partially immersed in the water at an angle. It seems bent at the point of immersion. The part seen out of the water is an effigy correctly placed on the object. The part seen in the water is an effigy placed nearer to the surface than the material oar is. The two parts are joined at the point of immersion. If the observer looks at the immersed oar while keeping the axes of his eyes vertical or nearly so, the bending effect that he sees is much smaller than if he looks at it from one side very obliquely.

**159.** To resume examining the various optical systems in their proper order, we should now take up the plate with plane parallel faces. But we shall treat it briefly because the subject is not very important.

Consider a plane parallel plate of thickness $e$, made of a material whose refractive index relative to the medium outside is $n$. When the plate is traversed by a spherical wave coming from a point source $S$, it lets the wave pass through, while reducing the amplitude of its vibration more or less, but deforms it. The emergent wave has a contour of the type that we previously found in the case of refraction through a single surface (§ 155). Its evolute is a caustic of revolution around the perpendicular drawn from the source to the faces of the plate. The cusp of the caustic is on this perpendicular and is displaced from the source by a distance $d$, given by

$$d = e \frac{n-1}{n}$$

For plates of ordinary glass, $d$ is equal in round numbers to about $1/3\ e$. Hence it becomes perceptible only when the plate is fairly thick and the observation is made at close range. When the observation occurs in a direction markedly oblique to the perpendicular drawn from the source to the faces, the situation grows complicated. For the wave segments that enter the observer's eyes are astigmatic. The more the direction of observation is inclined to the perpendicular, the farther the direction of propagation of the wave segments recedes from the cusp of the caustic.

In this connection, however, we may repeat what was said in § 157 about refraction through a single plane surface. If the mind took into account all the information reaching it by way of the eyes, it would arrive at very complex and almost unintelligible conclusions. But it explains away all the difficulties, fills up all the gaps, levels out the differences, and ignores whatever clearly refuses to become part of the over-all compromise. As a result, most of the time the effigy is created and placed as if the plate were not present.

Particularly if the object is extended and complicated as well as very familiar to the observer, its effigy is located mainly on the basis of data supplied by the memory and experiment. The plate may be said to exert practically no effect. Instances are rare in which the effigy created is definitely and perceptibly displaced from the position it would have occupied in the absence of the plate.

**160.** Let us now go on to consider the plate whose faces are plane and not parallel. The face upon which the waves are incident forms with the face from which they emerge an angle $\alpha$ different from zero. If $\alpha$ does not exceed three or four degrees, the plate is called a *wedge* or *thin prism*. If $\alpha$ amounts to several dozen degrees, the plate is known simply as a *prism*. However, to distinguish those prisms that act by refraction alone from others that function by reflection too, the former are sometimes labeled *refracting prisms*.

In general a prism exerts its effect on a plane wave in deviating it by an angle $\delta$ toward the base of the prism (§ 317). This angle $\delta$ is a function of the angle of incidence, of the prism angle $\alpha$, of the prism's refractive index $n$ (relative to the medium outside) and therefore by implication of the wave length $\lambda$.

This function becomes very simple in the case of thin prisms, because the effect of the angle of incidence may be regarded as negligible; then

$$\delta = (n - 1)\alpha$$

Thus if a plane wave impinges upon a wedge, it is deviated very simply. If the wedge is made of glass and is surrounded by air, the deviation is slightly more than half of the angle $\alpha$ (Fig. 57).

If the emergent wave reaches an eye, the segment entering it is rendered convergent and concentrated on the fovea, exactly as when the eye is affected by waves from a point source very far off. The customary physiologico-psychological process then occurs, at the end of which the mind creates a star effigy in the direction perpendicular to the waves received by the eye. If both eyes are involved, the information supplied by convergence confirms the remoteness of the source. The observer accordingly places the star effigy at the distance suggested to him by environmental circumstances, just as he does whenever plane waves strike his eyes.

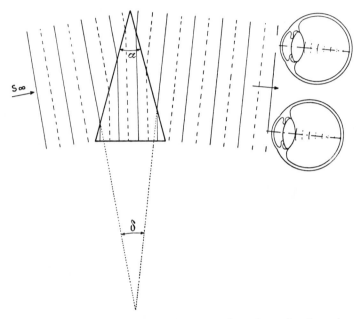

FIG. 57. Infinitely distant point source seen through a refracting prism

Yet the effigy is certainly not located where the real source is. For not only is the placement erroneous in depth, but the waves incident upon the prism came from a different direction.

**161.** Now suppose that $S$, a point source of waves, is at a finite distance $x$ from a thin prism's face of entrance (Fig. 58). Hence the waves incident upon the wedge are spherical. But since the deviation is not a function of the angle of incidence, every element of a wave is rotated through a uniform angle. Therefore, the emergent waves also are spherical, with their center at a point $I$ that is displaced, with respect to $S$, toward the edge of the prism. The displacement $s$ of $I$ with respect to $S$ is expressed by

$$s = x\delta = x(n - 1)\alpha$$

When the emergent waves are received by the eyes of an observer and analyzed, his mind is informed that these waves emanate from a point $I$, whose position is identified by the usual mechanisms of accommodation and convergence. The mind accordingly creates a star effigy and places it near $I$, except for errors in estimating the distance of the center $I$ from the eyes.

If instead of being a point the source has finite dimensions, we may regard it in our habitual way as the aggregate of its points. For each of these the reasoning set forth just above may be repeated. In the end the observer creates an assemblage of effigies like those

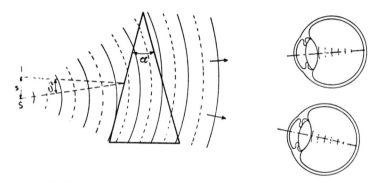

Fig. 58. Point source at a finite distance, as seen through a refracting prism

that he would have created in the absence of the prism. He stations these effigies at the same distance in front of himself, but they are all displaced by a length $s$ perpendicular to the direction of the prism's edge.

In this connection the following experiment is of interest. Place a wedge so that the emergent waves affect only one eye $O_s$, while the other eye $O_d$ receives the direct waves emitted by an object $S$ (Fig. 59). Let the eyes keep the attitude that they had while looking at their surroundings before the prism was introduced. But now, with the prism in front of one of the eyes, they see double, for the effigy created by the left eye is displaced by a length $s$ from the effigy created by the right eye. Whenever the ocular system is able to do so, however, it modifies itself in whatever way is necessary to avoid diplopia. Thus if the prism is in front of the left eye, as in Fig. 59, with the edge of the prism vertical and facing the nose, the convergence of that eye must be increased by an angle equal to the deviation δ of the prism in order to avoid diplopia.

As a result strange information reaches the mind. On the one hand, the strength of the convergence indicates that the distance between the object and the eyes is short. On the other hand accommodation, and in particular the data derived from experiment and the memory, guarantee that the object is still where it was prior to the introduction of the prism. Consequently the mind in the usual way places the effigy exactly as if the prism were not there.

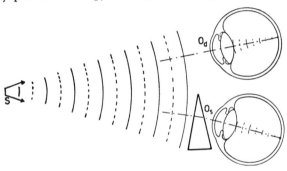

Fig. 59. Binocular vision of a source at a finite distance when a prism is placed in front of only one eye

The same thing happens if an equal or different prism with its edge oriented in any direction whatever is put in front of each eye. Unless the eyes succeed in achieving fusion, the observer sees double, the effigies created by the right eye being displaced with reference to those created by the left eye. But they are located at precisely the same distance from the eyes as though these had achieved perfect fusion. Then if fusion is accomplished by an appropriate rotation of the eyeballs, seeing occurs as if the prisms were not present; in other words, the information supplied by convergence is completely ignored by the mind. Yet some observers are deeply disturbed by these prisms that require the eyes to perform unusual rotations, and they prefer to close their eyes or look with only one eye.

**162.** What was said about thin prisms does not apply to refracting prisms of wide angle (60° is the type generally made) on account of various complications. Thus the variation of the deviation $\delta$ as a function of the wave length was imperceptible in the case of wedges, but now becomes the predominant phenomenon. The consequence is *dispersion;* that is, if the incident wave is complex, the various monochromatic waves that compose it undergo perceptibly different deviations. Then the eyes of the observer receive waves of many different inclinations (even if the source is a point). Each wave, after being rendered convergent by the optical system of the eye, affects a separate region of the retina.

Fig. 60 schematically illustrates the instance in which the incident wave consists of two waves of wave lengths $\lambda_1$ and $\lambda_2$. When

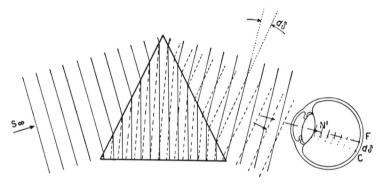

Fig. 60. Dispersion of waves through a refracting prism

these two waves emerge from the prism, they proceed in different directions, forming between them an angle $d\delta$ (§ 318). Suppose that the eye that receives them is oriented with its axis perpendicular to the wave $\lambda_1$. Both waves are made convergent. But the center of wave $\lambda_1$ affects the retina in the customary small group of cones at the center of the fovea $F$ (for the incident wave is assumed to be plane, and therefore also the emergent waves). The center of wave $\lambda_2$, on the other hand, stimulates a different group of cones at a point $C$, displaced from $F$ by an arc that subtends $d\delta$ at $N'$, the second nodal point of the eye.

The mind is thus informed that two plane waves have impinged upon the eye from two different directions, just as when in direct vision waves arrive from two very remote point sources. But now there is again another circumstance, the variation in $\lambda$. This produces different effects on the photosensitive material of the retina. For the impulses leaving from $F$ have a carrier frequency and in particular a modulated frequency different from those that depart from $C$. In the end the mind is led to create two effigies in the form of a star, one in the direction perpendicular to the wave $\lambda_1$, the other in the direction perpendicular to the wave $\lambda_2$. The former is endowed with a certain intensity and a certain hue; the latter, with a certain intensity, which may even be equal to that of the former effigy, and with a different hue. More exactly, most observers (those called *normal* from the colorimetric point of view) ascribe a hue closer to violet, in the order of the colors of the rainbow, to the more deviated wave.

At what distance from the eye will these two effigies be stationed? This is the usual problem. But now it is complicated by the scarcity, if not complete absence, of data derived from the memory and experiment. Hence the observer's imagination intervenes to unify (and sometimes even to modify) the optical information. Consequently the strangest results are obtained, from one observer to another, especially if he is a layman and entirely unfamiliar with the phenomenon. It may be said that no one ever does what is required by the optical information, namely, see two stars very far off. This experiment being generally performed in a darkened laboratory, the mind of the observer is induced to locate the effigies inside the room, even if the prism is big enough to let the experiment be conducted binocularly.

**163.** Suppose that the incident wave, instead of being composed of two monochromatic waves of wave length $\lambda_1$ and $\lambda_2$, contains three, four, . . . $m$ waves of wave length $\lambda_1, \lambda_2, \lambda_3, \ldots \lambda_m$. When these component waves emerge from the prism, each of them follows a direction of its own. If $\lambda$ decreases in the foregoing order, the deviation increases in the same order. When these waves enter an eye, they are made convergent and are concentrated each in its own group of cones on the retina. The result is the creation of a corresponding number of star effigies distributed along a straight line perpendicular to the edge of the prism. The various stars have different hues ranging, in the foregoing order, from red to violet according to the sequence of the colors in the rainbow. Thus the star closest to violet is displaced farthest toward the edge of the prism. The distance of this row of stars from the eye is subject to all the uncertainties mentioned in § 162.

The incident wave may be so complex that it contains all the visual waves, or at any rate all those comprised within a certain interval of wave length $\lambda_1 - \lambda_2$. Then innumerable waves emerge from the prism in as many different directions without any break in continuity, but they are all parallel to the prism's edge.

If they enter an eye, they are rendered convergent, and all the retinal elements along a stretch perpendicular to the edge of the prism are affected, just as when the eye receives directly the waves emitted by the points of a remote linear object. Hence the mind creates a linear effigy, oriented in the direction perpendicular to the prism's edge, and stationed at a distance that depends most of all on the environmental reports and the imagination of the observer.

But now there is a new fact concerning the linear object, for the retinal elements are affected by waves of continuously different $\lambda$. The effects are clear. Nerve impulses leave from each element with different frequencies. This leads the mind to modify the color continuously along the effigy, giving it a red hue in the part farthest removed from the prism's edge, and the various hues of the rainbow in turn up to violet in the part nearest to the edge. If the point source emitted all the visual waves, the effigy thus created by the mind is called a *continuous spectrum*. If, however, the waves emitted by the source are all those, and only those, included in the interval of wave length $\lambda_1 - \lambda_2$, then the created effigy will still be linear, but it will be only a part of the continuous spectrum.

**164.** Now suppose that the source is no longer a point but a linear object (for the present we may assume it to be very far away) oriented in a direction *parallel* to the prism's edge. If this source emits waves of a pure wave length $\lambda$, for each of its points the mind will repeat the process described in § 160. Hence the effigy created will be linear, parallel to the prism's edge, and displaced toward it by an angle equal to the deviation $\delta$ suffered by the waves.

A complication now appears, however. In our discussion of a point source we tacitly took for granted that the source and the center of the eye were in a single plane that was perpendicular to the edge of the prism and, therefore, cut the prism in a normal section. We regarded this plane as coincident with the plane of the paper, and we measured $\alpha$, the refracting angle of the prism, in it. But if the line joining the source to the center of the eye is not perpendicular to the prism's edge, then the waves are not parallel to the edge. Instead they form with it a certain angle $\varphi$. Hence the prism's effective section (if I may so call it) is no longer a normal section, but another section inclined to the normal section at the angle $\varphi$. The refracting angle is no longer $\alpha$, but $\alpha/\cos \varphi$. Therefore the deviation is greater.

As a result the line of affected retinal elements is not what it would be if the eye looked directly at a linear object in front of it. Instead the line now comes out curved and symmetrical with respect to the plane perpendicular to the prism's edge and passing through the eye. Consequently a curvilinear effigy is created that is symmetrical with respect to this plane and that always has its concave side facing the prism's edge.

For this curvature to be noticed by the observer, the angle $\varphi$ must not be too small, or the linear source must be long in comparison with its distance from the prism. But generally the object is very short, and then the effigy created by the mind is a segment of a straight line parallel to the edge of the prism. If the prism is rotated, however, so as to vary the angle $\varphi$, the segment effigy is seen inclined. Essentially it is a segment of the curved line mentioned above.

**165.** Suppose that the linear source (which we shall now assume to be short and parallel to the edge of the prism) emits waves of two, three, . . . $m$ different wave lengths. Then the final effigy is composed of a corresponding number of segments, parallel to the edge and therefore to one another. Their hue of course varies. In the order of the colors of the rainbow, the segment farthest from the edge has a hue closest to red, while the segment nearest to the edge has a hue closest to violet. This pattern is called a *line spectrum*.

If the linear source emits all the visual wave lengths, the effigy created is a streak consisting, as it were, of innumerable segments parallel to the edge of the prism. Each has a different hue, and all the colors of the rainbow are present, the violet being toward the edge of the prism. This pattern too is a *continuous spectrum*. It is made up, as it were, of innumerable linear spectra, like the one described in § 163 as produced by a point source. These spectra are situated one above the other, so that the points of like color form a segment parallel to the edge of the prism.

If the point source is not infinitely distant, the waves arriving at the prism's face of incidence are spherical, but are considerably deformed when leaving the face of emergence. The various elements of a spherical wave may be considered to be plane elements, each incident upon the prism at a different angle and therefore suffering a different deviation. The emergent wave, then, as an aggregate of elements that have undergone the action of the prism in different ways, can no longer have a spherical form. It nevertheless has a plane of symmetry, namely, the plane perpendicular to the prism's edge and passing through the source.

If this wave enters an eye, it is made convergent inside the eye. But the concentration on the retina may or may not be contained in the small group of cones in which it would have been contained had the wave received by the eye been spherical. Whether or not it is so contained depends, of course, on the extent to which the deformed wave emerging from the prism differs from a sphere. This deformation in turn depends on the angular amplitude of the wave segment traversing the prism and on the average angle of incidence. I am sticking to generalities because I do not intend to develop this topic in detail. For our purpose it is enough to conclude that if experimental conditions are such as to make perceptible the deformations of the waves emerging from the prism, then on the basis of the information furnished by the retina, the effigy created at the end of the mental process will be a small figure of strange form. The only statement that can be made about it is that it will have an axis of symmetry perpendicular to the edge of the prism. Under these conditions the placement of the effigy is influenced even less by the optical information, which is quite uncertain. We may say that the placement is based exclusively on data derived from the environment, memory (if there are any such) or even the imagination. There is a big difference between what someone says he sees who is observing these things for the first time, and what is reported by somebody who observes them with familiarity and knowledge, if not with preconceptions generated by scientific hypotheses or discussions.

**166.** We are now in a position to understand how complicated it must be to describe what *may* be seen by an observer looking through a refracting prism at nearby extended objects. The individual waves emitted by the points of these objects are spherical before reaching the prism, but emerge from it deformed. At the same time they are in every case modified by the value of $\lambda$ emitted by the corresponding element of the source. Each such wave undergoes a different deviation, a different deformation, and a different dispersion.

On the other hand, the waves emitted by an element of the object emerge from the prism superimposed upon those emitted by another element and deviated differently. This entire hodgepodge of waves arrives at the pupil and is transformed into an assemblage of convergent waves. These excite an enormous number of retinal elements. Along another enormous number of nerve fibers a corresponding quantity of trains of nerve impulses leaves, with different carrier frequencies, different modulated frequencies, and different indices of modulation. The mind has to analyze all this information and represent it in the effigy that it creates and must locate behind the prism. This is no mean task. Anyone who looks through a prism at diverse extended objects for the first time is apt to stare for minutes on end before being able to say what he sees; and to say what he sees means to describe the effigy created.

I do not intend to tarry longer over this subject, because it is enough to have made clear the criterion that should guide the reasoning of those who wish to account for what an observer sees when he looks through a refracting prism at complex objects emitting waves of many different wave lengths or even all the visual wave lengths.

**167.** To proceed in an orderly way in studying vision by means of optical systems, we should now go on to discuss refraction through a spherical surface separating two media of different refractive index. But since this is an optical system of little practical interest, we shall skip over it and go on at once to the most important system, the lens. This is a highly significant subject, which we shall have to develop step by step, even if a great deal of time is required.

Lenses may have various forms and various curvatures. First of all consider a lens having spherical surfaces, a circular rim, and a thickness negligible in comparison with the radii of curvature of the surfaces. Such a system possesses symmetry of revolution around a straight line, called the *optic axis*, which passes through the centers of curvature of the two surfaces.

Suppose that a point source $S$ emitting waves of a pure wave length $\lambda$ is situated on the axis of the lens. The spherical waves emitted by the source reach the lens, pass through it, and emerge in a form that may be very complex, although still possessing symmetry of revolution with respect to the optic axis (§ 320). For the sake of simplicity we shall assume this form to be spherical or plane. This supposition approaches the truth more closely, the smaller the angle subtended by the diameter of the lens at the centers of curvature of the surfaces.

The emergent wave has its center $I$ at a distance $x'$ from the center $O$ of the lens in accordance with the equation

$$\frac{1}{x} + \frac{1}{x'} = \frac{1}{f}$$

where $x$ is the distance from the source to the lens, and $f$ is the *focal length*, given by

$$\frac{1}{f} = (n - 1)\left(\frac{1}{r_1} + \frac{1}{r_2}\right)$$

where $n$ is the refractive index of the material of the lens, $r_1$ and $r_2$ being the radii of curvature of the two surfaces.

On the basis of these preliminary statements the analysis proceeds very much as in the case of spherical mirrors, a similarity that will let us use somewhat condensed language. Moreover, since lenses are rarely big enough to permit binocular vision, we may confine our investigation to monocular vision.

**168.** Let the source $S$ be at $F$, the *first principal focus* of the lens (Fig. 61). This focus is that point on the axis whose distance from

$O$, the center of the lens, is $f$. Then in the first equation of § 167, $x = f$, and therefore $x' = \infty$; that is, the emergent waves are plane and perpendicular to the optic axis.

If an eye $O'$ receives these waves, it makes them convergent and concentrates them at the fovea $F'$ in a tiny group of cones. The mind of the observer is thus informed that in front of the eye there is a point source of waves in the direction of the axis of the eye. The next step is to place the star effigy at a certain distance. But at what distance? This problem is difficult for the mind, because the only telemetric optical information is that furnished by accommodation, the slight value of which was indicated in §§ 98–100. The

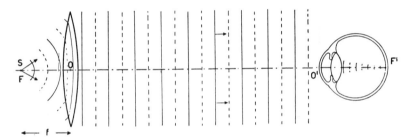

Fig. 61. Vision of a point source located at the focus of a converging lens ·

placement will consequently be based exclusively on reports of another kind.

If the observer knows where $S$ is, he will end up by locating the effigy exactly where the source is, or close by. What never happens is that the effigy is stationed at a very great distance behind the lens. This experiment is generally performed in a laboratory (the point source must be in complete darkness to attain the assumed conditions) whose dimensions are known to the observer. It will never enter his mind to think of the walls as being perforated and thereby making a remote star visible. Hence the effigy will always be located a short distance away. The precise value varies a good deal from one observer to another, and depends on the particular circumstances in which the experiment is conducted.

**169.** Now let us bring the source closer to the lens, so that $x < f$ (Fig. 62). This time the wave emerging from the lens is divergent, with its center at a point $I$ behind the lens at a distance $x'$, given by the first equation in § 167 and therefore always numerically greater than $x$.

The waves arriving at the eye $O'$ are once more made convergent by accommodation of the crystalline lens, and are concentrated at the fovea $F'$ in the usual small group of cones. The mind must create a star effigy and locate it on the axis of the eye and of the lens. But at what distance? The conditions are practically unchanged from those in § 168. The curvature of the wave requires only a slight

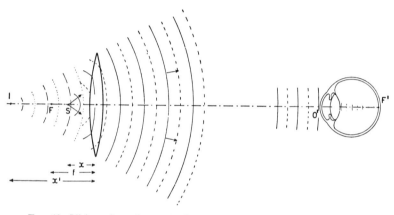

Fig. 62. Vision of a point source located between a converging lens and its focus

accommodation, which in general exerts no influence on the decisions of the mind. Most of the time the effigy $I$ is placed where the source $S$ is, or nearby, especially if the observer has ways of identifying the position of the source, for example, by means of touch.

Let us now turn back and move the source $S$ away from the lens until it is a little beyond $F$. For the purpose of describing the most interesting effects more easily, let us again take up a specific case. Let $f = 50$ cm, while the distance from the eye $O'$ to the lens $O$ is 2.5 m. The conditions chosen are the same as those in § 138 for a concave spherical mirror, not only because we thus avoid new calculations but also because a comparison of what happens in the case of a mirror with what happens in the case of a lens will serve to clarify the general mechanism under discussion.

Let us then move the source $S$ to 54 cm from the lens (Fig. 63). The emergent wave is concave, and its center is at a point $I$ 6.75 m from the lens on the same side as the eye $O'$. This, therefore, receives convergent waves, but since their vergence does not attain 1/4 $D$, it

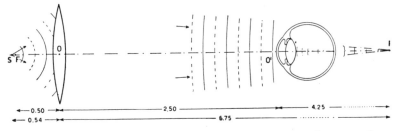

FIG. 63. Point source seen through a converging lens, when the center of the waves emerging from the lens falls behind the eye of the observer

does not produce any perceptible effects. Hence the star effigy will be placed on the axis, just as when the source $S$ was at the focus $F$.

**170.** Let us now move $S$ farther away from the focus until it is 60 cm from the lens. The emergent waves are still convergent, with their center $I$ 3 m from the lens or 50 cm behind the eye. If this is normal, the center of the wave within the eye falls somewhat in front of the retina, and to invoke accommodation only results in pushing this center farther forward. Therefore, the number of stimulated retinal cones exceeds the minimum, and the effigy created is larger than the one mentioned at the end of § 169. Hereupon a strange situation arises. For the mind no longer has any optical data to localize this cumbersome effigy, which moreover loses all affinity with the source, since the latter has the form of a point.

In the habitual way the effigy is attached to the first object of which the mind has knowledge along the path of the waves, and this is regularly the lens. The observer sees a roundish luminous spot in the middle of the lens. This spot expands as the source $S$ is moved away from $F$ until it attains the distance of 62.5 cm from the lens $O$. The spot then occupies the entire lens; that is, the whole lens appears luminous to the observer.

The reason is clear. When the source is at this distance from the lens, the waves emerging from the lens have their center $I$ exactly in the pupil of the eye $O'$ (Fig. 64). Hence as they penetrate the eye, they diverge and on the retina affect a circular region whose angular aperture is precisely equal to that of the lens. The effigy accordingly will be a luminous disk, which will be located on the lens as the first material object in front of the eye. Thus the lens will appear all lit up.

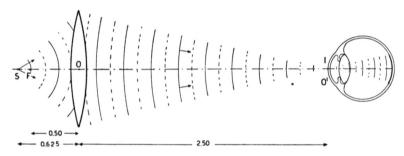

FIG. 64. Point source seen through a converging lens, when the center of the waves emerging from the lens falls on the eye of the observer

Let us continue to move $S$ farther away from the focus $F$ until it is 67 cm from the lens $O$ (Fig. 65). The waves emerging from the lens are even more convergent, their center $I$ being 2 m from the lens or 50 cm in front of the eye. They reach the eye $O'$ divergent, and with $2D$ of accommodation are concentrated in the minimum group of cones in the fovea $F'$. Consequently the mind must again create a star effigy and place it on the axis of the eye, which is also the axis of the lens. The familiar question pops up: at what distance? If accommodation counted for something, the star should be placed about 50 cm from the eye. But this is as usual a decision in conflict with common sense. Unless the mind has been subjected to special training, it refuses to accept any such decision. Since a luminous star motionless in mid-air a couple of feet away from the eye is an absurdity, the mind projects the effigy upon the first material object in the path of the waves, and customarily locates it on the lens. But if some material object, like a perforated diaphragm or a support, is put near the center $I$, then the effigy may be situated at that point without any difficulty.

**171.** If we continue to move the source $S$ farther away from the lens, the center $I$ approaches the lens and recedes from the eye. The effort of accommodation needed to reduce to a minimum the group of affected retinal elements constantly decreases. The effigy is still created in the form of a star and is placed on the optic axis. So far as its distance from the eye is concerned, it is usually located on the lens, unless objects apt to induce the mind to station the star else-where are put between the lens and the eye.

The mind may of course be convinced by other means, for example, by the power of persuasion. Since the arrangement con-sidered thus far is frequently used in optical laboratories, those who

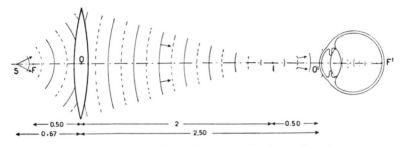

Fig. 65. Point source seen through a converging lens, when the center of the waves emerging from the lens falls in front of the eye

have acquired familiarity with it do not find it difficult to station the star in mid-air a couple of feet away from the eyes, where they know by other indications that the center of the waves must be.

In this connection the application of temporal parallax is inter-esting. In Fig. 65 move the eye in the direction perpendicular to the axis of the lens, while keeping the pupil on the wave front whose center is $I$. As was made clear in regard to direct vision (§ 110), the mind has a way of obtaining from this movement, by evaluating the resulting change in the retinal elements affected, information about the distance from the eye to the center of the waves received.

From this process the mind should infer that the center $I$ is a couple of feet away from the eye, and should locate the star effigy accordingly. But if the preconception prevails that the star cannot be there, and the effigy is consequently attached to the lens, then temporal parallax too does not succeed in avoiding this placement.

But a new effect is noted. When the eye is moved upward, the stimulated retinal elements are no longer at the center of the fovea, but higher; hence the effigy is located lower on the lens. Contrariwise, if the eye is lowered, the effigy moves upward on the lens. In other words, the observer sees the star displaced on the lens with a motion parallel to that of the eye but in the opposite direction, a motion that may be called *inverse*. When the pupil arrives at the edge of the wave, the effigy reaches the rim of the lens. If the pupil goes beyond the edge of the wave, obviously the eye no longer receives any radiation and as a result the mind no longer creates an effigy. During this movement the observer has the impression that the star *is setting* behind the rim of the lens.

This course taken by the effigy does not offer any unacceptable absurdities to the observer. It is a much less indigestible morsel to swallow than the presence of an unsupported point source motionless in mid-air 50 cm from the eyes. Therefore, temporal parallax serves no useful purpose, unless of course the observer is a trained initiate.

**172.** Binocular vision, when it can be applied, has the same effect as temporal parallax. Thus under the conditions depicted in Fig. 65 (it would be even better if the source $S$ were moved still farther away from the lens, or the observer backed up somewhat) the head may be placed so that the wave with its center at $I$ enters both eyes at the same time. To avoid diplopia, the eyes must converge on $I$. But that conflicts with the conviction of the mind, which wants to locate the effigy on the lens $O$. The eyes are made to converge on $O$, but then two stars are seen.

A state of discomfort ensues. The observer shuts his eyes completely, or closes one of them, or moves aside so as to receive the waves in only one eye, or looks away. Occasionally he even succeeds in persuading himself to station the star effigy at $I$.

Various other observations could be made with this arrangement. Some might even be useful in practice because, as was mentioned in § 171, a setup of this kind is often utilized in optical laboratories. For example, in the circumstances schematically illustrated in Fig. 63, if the eye is displaced in a direction perpendicular to the axis, the star effigy that had been located on, or a little behind, the lens is seen to move on the lens with a motion *conforming* to that of the eye, the star being seen to set behind the rim of the lens when the pupil goes beyond the edge of the wave. But these appearances, which are easily explained by the mechanism expounded in the preceding pages, need not be scrutinized in detail.

**173.** Let us instead go on to examine the case of extended sources.

A point source $S_1$ that projects waves upon a lens may be, not on the axis, but at a distance $y$ from it (Fig. 66). Then the waves

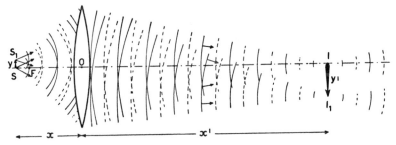

FIG. 66. Waves emitted by the elements of an extended source undergoing modification of their course when passing through a converging lens

emerging from the lens possess, not symmetry of revolution about an axis, but only a plane of symmetry. Yet when the straight line joining the source $S_1$ to $O$, the center of the lens, cuts the optic axis of the lens at a small enough angle, the differences between the emergent wave and a comparable sphere are imperceptible. It may, therefore, be said that the waves emitted by the source $S_1$, after passing through the lens, emerge sensibly spherical again. Their center is at a point $I_1$ on the straight line $S_1O$. The distance $x'$ between $I_1$ and $O$ is still given by the first equation in § 167, $x$ being the distance $S_1O$.

A linear object $SS_1$ (which we shall consider perpendicular to the axis of the lens) may be regarded as an aggregate of point elements. Each of these on its own account emits waves, which do not disturb one another. The lens makes those emitted by $S$ spherical with their center at $I$. Those emitted by $S_1$ are made spherical with their center at $I_1$. Those emitted by any point between $S$ and $S_1$ are made spherical with their center at a point between $I$ and $I_1$; but the straight line joining this center with the corresponding emitting element must pass through $O$. Thus between $I$ and $I_1$ there is a whole series of centers of waves, and these centers constitute a segment whose length $y'$ is obviously related to $y$, the length of the segment $SS_1$, by the equation

$$\frac{y'}{y} = \frac{x'}{x}$$

**174.** After these preliminary remarks consider an object $SS_1$, whose extremity $S$ coincides with $F$, the first principal focus of a

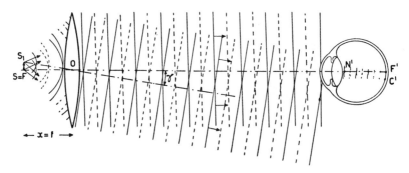

FIG. 67. Extended source seen through a converging lens, when the source is situated at the focus of the lens

lens (Fig. 67). When the waves emitted by each point emerge from the lens, they are plane. In accordance with our usual practice we shall examine only the waves emitted by the extremities, since those in between follow an intermediate course that may be easily determined.

If this mass of plane waves impinges upon an eye, a portion enters and is made convergent. Each of these waves affects a small group of retinal cones. Thus there is a whole row of stimulated elements on the retina between $F'$ and $C'$. $F'$ in the center of the fovea is the terminus of the waves emitted by $S$, while $C'$ is the terminus of those emitted by $S_1$. The corresponding impulses transmitted to the brain by way of the optic nerve inform the mind

that in front of the eye there is a linear object, whose lower extremity
is in the direction $N'F'$ and whose upper extremity is in the direction
$N'C'$; in other words, its angular magnitude $\gamma$ is equal to that
subtended by the object $SS_1$ at $O$, the center of the lens.

As a result of this report the mind must create a linear effigy.
But at what distance should it be placed? If the information obtained through accommodation carried any weight, the mind would
have to station an enormous wire very far away. But we have
already noticed on many occasions how little account is taken of
the data delivered by accommodation. Instead the effigy is localized
in accordance with utterly different criteria. These are affected in
the main by environmental conditions or by the knowledge the
observer has of the position and nature of the object. I have already
in many cases indicated the dominant influence exercised by these
criteria in the placement of effigies.

**175.** As usual, unless special precautions are taken, the observer knows where the object is, either because he actually holds it
in his hand or because it rests on a table or support that he sees
behind the lens. In very many instances it is these circumstances
that make the mind decide to place the effigy in the plane of the
object.

If this happens (Fig. 68), $y_e$, the length of the effigy, is given by

$$y_e = \gamma(d + f)$$

where $d$ is the distance from the eye to the lens. But

$$\gamma = \frac{y}{f}$$

and therefore

$$\frac{y_e}{y} = \frac{d + f}{f}$$

This equation, by indicating how many times greater than the
material object the effigy is, thereby gives us the magnification of
the lens under those conditions and on the hypothesis that the
effigy is placed in the plane of the object.

Numerous observations and measurements carried out for the purpose of testing the reliability of this equation have pretty generally confirmed it. Yet $d$, the distance from the eye to the lens, plainly enters into it, and therefore we cannot speak of the absolute magnification of a lens, even assuming the foregoing hypothesis to be true in every case. In particular it is interesting to note that the

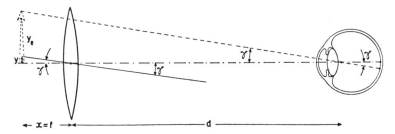

Fig. 68. Magnification of the figure seen through a converging lens, when the object is at the focus of the lens

magnification increases as the eye recedes from the lens, the minimum value being unity when $d = 0$, for then

$$\frac{y_e}{y} = \frac{f}{f} = 1$$

We thus obtain the rule, which may be verified at once, that "when the lens is in contact with the eye, objects located in the focal plane seem to be of the same size as they would be without the lens." It is not easy, however, to avoid having some distance, even if small, between the lens and the eye. This distance must be negligible in comparison with the focal length of the lens, if the experiment is to confirm the rule. The decisive tests will be discussed in § 193, however, because the condition that the object must be in the focal plane is excessively and uselessly restrictive. Yet this condition may be useful in some verifications. For instance, if we put $d = kf$, where $k$ is any positive integer, it follows at once that

$$\frac{y_e}{y} = k + 1$$

In other words, for those distances of the eye from the lens, the magnification is independent of the lens' focal length. This behavior too is experimentally confirmed, so long as the hypothesis is valid that the effigy is placed in the plane of the object. This placement is of course not always possible, as for example when the focal length of the lens is just a few inches.

**176.** Now suppose the object approaches the lens a little. As we saw in § 169 with regard to a point source, the present situation differs from the preceding one, in which the object was in the focal plane of the lens, only because this time the waves emerging from the lens are more or less convex, with their convexity increasing as the object moves farther away from the focus toward the lens.

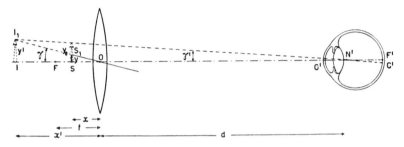

FIG. 69. Magnification through a converging lens, when the object is at any distance from the lens

Hence the mind must again create a linear effigy oriented like the object. The effigy's angular magnitude is determined by the length of the retinal trace, while its placement is decided once more by data derived from the environment and the memory. Accommodation, although carrying small weight, furnishes reports always favorable to putting the effigy in the near vicinity. It generally happens that the effigy is placed in the plane of the object.

If this hypothesis is verified, the magnification, which was defined in § 175 as the ratio between $y_e$, the length of the effigy, and $y$, the length of the object, may be computed. The factors needed for the calculation are represented in Fig. 69, where the letters designate the following items considered in previous Figures: $SS_1$ is the linear object perpendicular to the optic axis of the lens; $y$ is the length of the object; its distance from the lens is $x$; the angle which it subtends at $O$, the center of the lens, is $\gamma$; $d$ is the distance from the eye $O'$ to the lens; $II_1$ is the locus of the centers of the waves emerging from the lens; $x'$ is the distance from the lens to the locus; its length is $y'$; the angle which it subtends at $N'$ is $\gamma'$; and $y_e$ is the length of the effigy located on the object. The interesting relations between the various magnitudes under discussion are the following:

$$\frac{y'}{y} = \frac{x'}{x}; \qquad \frac{y'}{y_e} = \frac{x' + d}{x + d}; \qquad \frac{1}{x} + \frac{1}{x'} = \frac{1}{f}$$

By eliminating $y'$ and $x'$ from these equations we obtain the formula

$$\frac{y_e}{y} = \frac{\dfrac{1}{d} + \dfrac{1}{x}}{\dfrac{1}{f} + \dfrac{1}{d} - \dfrac{1}{x}}$$

When subjected to experimental control, this formula has shown itself to answer the purpose reasonably well. This means that the hypothesis on which it is based, namely, that the effigy is placed where the object is, is constantly verified in practice. The verification is highly satisfactory so long as the focal length of the lens is not too small. It ceases to be valid when $1/f$ exceeds $20D$ or when the observer is prevented by suitable screens from seeing where the object is.

The regular placement of the effigy in the plane of the object is demonstrated, moreover, by the common experience of those who use a magnifying glass to observe any object more conveniently and more effectively. They "see the object bigger," but they see it neither nearer nor farther away. This statement may be translated into our language as follows: the effigy created as a result of the stimulus produced by the waves emerging from the lens is located where the object is.

**177.** To go on examining the effect of the lens under consideration, let us now suppose that the object $SS_1$ is moved away from the lens past the focus $F$ until $x$ is 54 cm, as was done with the point source in Fig. 63.

The waves emerging from the lens this time are convergent, with their centers $I \ldots I_1$ behind the observer. But on the retina things proceed as in the previous cases. Again the effigy is a wire oriented like the object and, when placed in the plane of the object, it comes out larger. The magnification, still considered as the ratio between the length of the effigy and that of the object, may once more be calculated by the formula given in § 176. It is now a little bigger than before. If the observer looks uninterruptedly through the lens while the object recedes, he sees a steadily enlarging figure because the effigy increases in length.

As this displacement is extended, the effigy continues to lengthen, but at the same time it begins to widen and become indefinite at the edges. The reason is clear. As the object is moved away from the lens, $x$ increases and, therefore, $x'$ must diminish. The centers of the waves emerging from the lens approach the cornea of the eye $O'$. The waves that enter are so strongly curved that despite the absence of accommodation the centers of the waves inside the eye fall in front of the retina and farther and farther from it. Consequently, for every point of the object there is a constantly larger group of affected elements on the retina. Hence the effigy created must be spread out and confused.

**178.** When the object reaches 62.5 cm from the lens, as happened to the point source in Fig. 64, the effigy becomes so big that it fills the entire lens. Under these conditions $x' = d$, because the centers of the emergent waves are exactly as far from the lens as is the eye $O'$. Since

$$\frac{1}{x} + \frac{1}{x'} = \frac{1}{f}$$

if we apply the appropriate signs

$$\frac{1}{f} + \frac{1}{d} - \frac{1}{x} = 0$$

the formula for magnification gives

$$\frac{y_e}{y} = \infty$$

The magnification is infinite. This is an expression of the fact, already mentioned, that the whole lens appears luminous, even if the object is very small or a mere point.

As we continue to move the object farther away from the lens, the centers of the waves emerging from the lens fall in front of the eye. Hence markedly divergent waves reach the cornea. Despite a strong effort of accommodation the center of the convergent waves inside the eye cannot be brought to the retina, but remains behind it. To every point of the object, therefore, a conspicuous group of affected elements corresponds on the retina. Accordingly the effigy will be widened and indefinite. In addition it is inverted. In other words, as the object recedes from the lens, the effigy contracts and shrinks. But where will it be placed?

The answer is familiar. According to the information furnished by accommodation, and possibly by convergence and temporal parallax too, the effigy should be located in mid-air somewhere between the lens and the eye. But the mind generally rejects this inference, and the placement is made on the basis of data derived from the environment and the memory.

**179.** How powerful these influences are may be shown by the following experiment. In full daylight the shutter outside an open window was closed. Near the window was placed a lighted electric lamp, arranged so that the bulb pointed up. A big lens having a focal length of 1 m was put 2 m from the shutter, with the observer 2.5 m from the lens. When looking at the lens, he should have located the effigies 50 cm in front of his eyes. Although the observers were experts in the subject, the results were uniform. They all located the effigy of the lamp, bulb downward, exactly 1/2 m in front of their eyes; and they placed the effigy of the shutter behind the lens, very near the actual shutter, even though the effigy was a shutter upside down.

We have now completed our discussion of the phenomena that appear when an observer looks through a converging lens. I do not pretend to have dealt with all the appearances found under these conditions. Rather, I have confined myself to the most important, and in particular to those more likely to familiarize the reader with the scheme of thought that I find suitable for treating these phenomena. Once this scheme has been mastered, it is fairly easy to account for the appearances that may show up under circumstances different from those considered here.

**180.** Let us go on then to analyze the behavior of a diverging lens (§ 322). The possible cases are numerous but fairly uniform. The effect of a diverging lens on plane or divergent spherical waves reaching it from a monochromatic point source is to increase their divergence. In other words, it transforms them into other spherical waves (or at any rate we may regard them as spherical, ignoring possible deformations) whose center is nearer to the lens than is the original center. To be more precise, if $x$ is the distance from the source to the lens (considered as of negligible thickness) and $x'$ is the

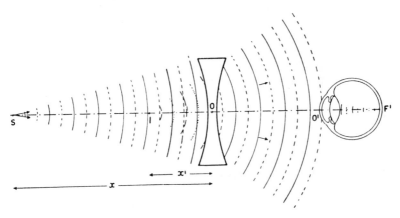

Fig. 70. Point source seen through a diverging lens

distance from the lens to the center of the waves emerging from it, again we have

$$\frac{1}{x} + \frac{1}{x'} = \frac{1}{f}$$

where the focal length is now taken as negative to show that the lens is diverging.

In Fig. 70 the source $S$, a point situated on the axis of the lens $O$, emits spherical waves that emerge from the lens as other spherical waves with their center at a point $I$ on the same side of the lens as the object. If an eye $O'$ receives these waves, it makes them convergent and, by a suitable effort of accommodation, concentrates them in a small group of cones in the fovea $F'$. The result is the creation of a star effigy, located on the axis of the eye and of the lens at a distance which, despite the effort of accommodation, may even be rather big because it is generally determined by data derived from the environment and the memory.

When the source, instead of being a point, consists of a segment $SS_1$, for each of its points we may repeat what was said for the source $S$ in Fig. 70. Let us also recall that, with the simplifications adopted in § 173, the waves emerging from the lens and corresponding to those emitted by a certain point of the source have their center collinear with that point and with $O$, the center of the lens.

Let the object under consideration, then, be a segment $SS_1$ perpendicular to the axis (Fig. 71). Let the length of the segment be $y$,

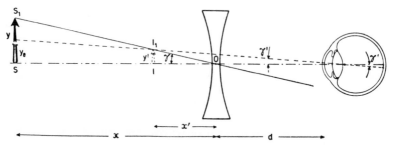

FIG. 71. Extended object seen through a diverging lens

and $x$ its distance from $O$. At distance $x'$ from the lens the centers of the waves emerging from the lens will be grouped at $II_1$, $I$ corresponding to $S$ and $I_1$ to $S_1$. If $d$ is the distance from the eye to the lens, the segment $II_1$ of length $y'$ subtends at the eye the angle

$$\gamma' = \frac{x' + d}{y'}$$

This is, therefore, the angular magnitude of the effigy created by the mind. We have also

$$\frac{y}{x} = \frac{y'}{x'} \quad \text{and} \quad \frac{1}{x} + \frac{1}{x'} = \frac{1}{f}$$

This set of relations resembles that found in § 176 for a converging lens, and the conclusions likewise are similar.

If the data derived from the environment and the memory again induce the mind of the observer to place the effigy where the object is, the effigy acquires a length $y_e$ connected with $y'$ by the equation

$$\frac{y_e}{y'} = \frac{x + d}{x' + d}$$

It is at once obvious that $y_e$ is smaller than $y$, this ratio decreasing as $d$, the distance from the lens to the eye, increases. Therefore a diverging lens *produces a diminution*, inasmuch as it leads to the creation of effigies smaller than the corresponding objects.

If $d$ is eliminated, that is, if the eye is brought into contact with the lens (accommodation being assumed to be adequate) the diminution is zero. This conclusion, which applies also to magnification by converging lenses (§ 175), is resoundingly verified in millions upon millions of cases by all those who wear eyeglasses.

**181.** We could now remove various limitations imposed on the preceding analysis, such as the monochromaticity of the radiation emitted by the source, and the sphericity of the waves emerging from the lens. But to do so would compel us to enter the complicated mazes of the mechanism of waves, and might make us lose sight of the thread of our argument. Hence I shall confine this Chapter to a discussion of the indispensable topics.

We come then to the consideration of an optical system composed of two lenses. We shall fix our attention on the most important of such systems, the telescope and the microscope. We shall of course treat these instruments schematically, ignoring all the intricacies due to the fact that the waves are not exactly plane or spherical.

The telescope consists of two lens systems, the *objective* and the *eyepiece*. The objective is a converging system having a very long focal length and may be represented by a simple converging thin lens. The eyepiece is either a converging or a diverging system, with a much shorter focal length than that of the objective. Since our reasoning does not change substantially whether the eyepiece is converging or diverging, we shall deal only with the former.

In the usual arrangement the two lenses are placed coaxially so that the focus of the eyepiece coincides with the center of the waves emerging from the objective.

**182.** Let a point source $S$ be on the axis of the system at a very great distance $x$ from the center of the objective $O$ (Fig. 72). The value of $x$ is of no particular interest. It is generally considered infinite, but need not be so at all. In any case, the waves emerging from the objective are convergent with their center at a point $I$. Thereafter they diverge and pass through the eyepiece $E$. Since $F_e$, the focus of the eyepiece, is at $I$, the waves emerging from the eyepiece are plane. In their path the eye of the observer is placed. Without any effort of accommodation the eye makes the waves convergent with their center at the fovea $F'$, where the usual small group of cones is therefore stimulated. By the customary transmission of nerve impulses the mind is informed that on the axis of

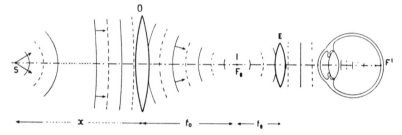

FIG. 72. Optical diagram of the telescope

the eye there is a point source of waves. Hence the mind must represent it by creating the habitual star effigy. This has to be located on the axis of the eye, which coincides with the axis of the optical system under the conditions of Fig. 72. But now the familiar embarrassing question arises: at what distance should the effigy be placed?

If accommodation counted for anything, the effigy should be located at infinity, a context in which we know well that accommodation is utterly unreliable. In these circumstances it is absolutely impossible to appeal to binocular vision, because the wave front is too small. Nor is resort to temporal parallax feasible, for the same reason. Therefore the mind can base a decision about the placement of the effigy only on data drawn from the memory and the environment, when they are not entirely products of the imagination. The strangest results are recorded. One person, guided by his knowledge of the position of the source $S$, puts the star effigy near the source. A second person, impressed by having before his eyes an instrument often equipped with a massive metallic mounting, stations the star inside the instrument. Others locate the effigy in intermediate positions. The outcome may be extremely varied, for an obvious reason.

**183.** Nevertheless we should continue to analyze the optical properties of this system. For this purpose suppose the source is displaced from the axis of the objective by being moved to $S_1$ (Fig. 73). Under these conditions it is useful to consider, not the distance from $S_1$ to the axis, but the angle $\gamma$ subtended by that distance at the center $O$ of the objective. The waves emerging from the objective

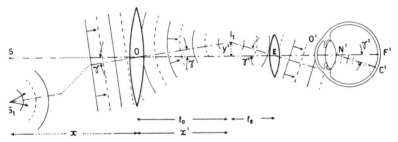

FIG. 73. Magnification in a telescope

have their center at a point $I_1$, which therefore is also outside the axis of the objective, on the side opposite $S_1$ with respect to the axis. The direction $OI_1$ and the axis form the angle $\gamma$ (§ 321). When the waves pass beyond $I_1$, they diverge and encounter the eyepiece $E$. This makes them plane, because they must be perpendicular to the direction $I_1E$, which joins the center of the waves with the center of the eyepiece. When they enter the eye $O'$, they are made convergent. But their center, while still on the retina, is no longer on the axis of the eye at $F'$, but at a point $C'$ on the side opposite $I_1$ with respect to the axis of the entire system. The distance $F'C'$ subtends at the nodal point $N'$ an angle $\gamma'$, which is exactly equal to the angle formed by the direction $I_1E$ with the axis.

The mind now finds itself in the position of creating another star effigy, which it places at whatever distance it pleases. But what is interesting to observe at this time is the direction in which the effigy is placed. This direction must form the angle $\gamma'$ with the axis, and must also be on the side opposite $S_1$. In other words, when $S_1$ starts from $S$ on the axis and is displaced by $\gamma$, for example, toward the right, the effigy is displaced by $\gamma'$ toward the left.

The distance from $I_1$ to $O$ is practically equal to $f_o$, the focal length of the objective, since we have assumed the sources $S$ and $S_1$ to be remote from $O$. The distance from $I_1$ to $E$ is practically equal to $f_e$, the focal length of the eyepiece. If $y'$ is the distance from $I_1$ to the axis, we have at once

$$\frac{y'}{f_o} = \gamma; \qquad \frac{y'}{f_e} = \gamma'; \qquad \text{and therefore} \qquad \frac{\gamma'}{\gamma} = \frac{f_o}{f_e}$$

**184.** Now suppose that the two sources $S$ and $S_1$, as well as innumerable other point sources between these two, exist at the same time so as to form a linear object $SS_1$, perpendicular to the axis of the system. For each of these sources we may repeat the analysis already made for $S_1$. Hence in the interval between $I_1$ and the axis we shall have countless centers of waves in a straight line, each of these centers being also in a line with $O$ and with the corresponding source. Finally we have between $F'$ and $C'$ a whole row of retinal elements affected by waves that may even be of different $\lambda$. But, ignoring the question of colorimetry, we must conclude that the information received by way of the optic nerve induces the mind to create a linear effigy perpendicular to the axis of the eye, and to place it wherever will be deemed best.

It should be observed that new data drawn from the memory and the environment now intervene because the figure of the effigy may be familiar to the observer, and this knowledge is not without effect on the placement of the effigy. We have already established that the angular magnitude of the effigy is $\gamma'$, while that of the object $SS_1$ is $\gamma$. The latter is measured from the center $O$ of the objective, but the former from the eye $O'$. Since the distance from the object to the objective has been assumed to be very great, in comparison with it the length of the telescope, that is, the distance $OO'$, may be ignored. The angular magnitude of the effigy may therefore be said to be $\gamma'/\gamma$ times, or $f_o/f_e$ times, greater than the angular magnitude of the object.

Suppose the observer's eye receives waves directly from the

object, and then receives them by way of the telescope. In the former case a small effigy is constructed; in the latter case, a bigger effigy, which is also inverted. We may now be asked, how many times larger than the first effigy is the second? This is a very thorny question, because the answer depends on the criterion governing the mind's placement of the two effigies. If they were stationed at the same distance, the two effigies would be to each other in the ratio $\gamma'/\gamma$. But this is not always the case. The second effigy may be located nearer, and then it is diminished in proportion. Certain observers even place the effigy created with the waves that pass through the telescope so near that this effigy has the same dimensions as the effigy seen in direct vision.

This effect is indicated by the following familiar expressions: this telescope *magnifies* so many times; this telescope makes objects appear so many times *nearer*. Obviously it is not the telescope that magnifies or makes objects appear nearer, but it is the observer's mind that utilizes as it thinks best the increase of angular magnitude produced by the telescope. Of course the results are very different, depending on whether the observer is dealing with familiar objects and whether the telescope increases the angular magnitude several times or many dozens or hundreds of times. In the case of small instruments for which $\gamma'/\gamma$ is 2 or 3, not a few observers find that they see the same with the instrument as without it. Clearly the mind at once brings the effigies back to the ordinary dimensions, especially if the objects are situated in the area fairly close to the eye, where the size of the effigies is estimated without giving much weight to the angular magnitude (§ 121).

In general it may be said that when the eye looks through a telescope, the mind makes the effigy a little bigger than the object and places it a little nearer, so as to effect a sort of compromise between magnification and approach.

**185.** Let us now turn back to observe through a telescope a point source $S$ on the axis of the system, as in Fig. 72, but with the eyepiece brought a bit nearer to the objective (Fig. 74). For the sake of simplicity, although the matter has no importance, suppose the source $S$ is infinitely distant, so that the waves arriving at the objective are plane and perpendicular to the axis. After passing through the objective, they become spherical and convergent, with their center at $F_o$. Then they diverge, pass through the eyepiece, and emerge still divergent with their center at a point $I$, which is farther away from the eye, the smaller the distance between $F_o$, the center of the waves, and $F_e$, the focus of the eyepiece. In other words, if the observer starts with the conditions in Fig. 72 and moves the eyepiece toward the objective, the center of the waves emerging from the eyepiece is displaced from infinity toward the eye. Since the eyepiece has a short focal length, the displacement is extremely rapid. For instance, an eyepiece having a focal length $f_e = 30$ mm need be brought only 1 mm closer to the objective for the center of the emerging waves to attain a distance $IO'$ of 87 cm.

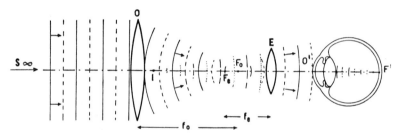

Fig. 74. Vision through a telescope from which divergent waves emerge

The emerging waves enter the eye and are made convergent. By accommodation of the crystalline lens their center is brought to the retina at $F'$. The only difference from the case of Fig. 72 is that now the crystalline lens accommodates as much as is necessary to compensate for the vergence of the wave arriving at the eye. Thus in the particular instance mentioned above it would accommodate a little more than $1D$.

If this effort counted for something, the mind, when informed of it, would have to station the star effigy on the axis of the eye about 1 m away. But since this information exerts no influence, the mind finds itself in the same situation as when $F_o$ coincided with $F_e$, and the result is the same. A star effigy is stationed on the axis of the eye at the same distance as before. For apart from accommodation, the factors that decide the placement are not altered at all.

Hence if the eyepiece is brought nearer to the objective while the observer is looking inside the telescope, the emerging waves are curved; from infinity their center moves to within a foot or two from the eye; the crystalline lens accordingly accommodates as much as is necessary; and the effigy remains motionless where it was. The observer declares that there is no change in what he sees.

That is true up to a certain point. If the eyepiece is pushed too close to the objective, the waves emerging from the former are so divergent that even with the greatest effort of which the crystalline lens is capable, it does not succeed in compensating for their vergence. The center of the waves inside the eye does not fall on the retina, but remains behind it. On the retina the number of affected cones exceeds the minimum, and the effigy comes out widened. For example, if the eyepiece has a focal length of 30 mm and is brought 5 mm closer to the objective, $6.7D$ of accommodation are required to compensate for the vergence of the emerging wave. A normal observer forty-five years of age no longer possesses so great a power of accommodation.

Thus are created the conditions by which an extended source gives rise to a widened effigy with indefinite edges. The observer says that he sees things *out of focus*, but the placement remains unchanged. By moving the eyepiece, he modifies the clarity of the outlines of the effigy, but he never varies its distance from the observer, that is, its placement in depth.

**186.** Similarly, if the eyepiece is moved away from the objective past the position characterized by the coincidence of $I$ with $F_e$, as in Fig. 72, the wave emerging from the eyepiece is convergent, with its center of curvature behind the observer's head. This center is nearer to the eyepiece, the farther the eyepiece is from the objective. Accommodation can no longer accomplish anything, for the waves entering the eye are convergent with their center at a point in front of the retina. This point is farther from the retina, the farther away the eyepiece is from the objective. The result is well known. The effigy widens and, in the case of an extended object, appears confused with indefinite edges. The observer says that he sees things *out of focus*, but the placement is unchanged.

Telescopes usually have an eyepiece movable in depth, whose position is varied with complete ease and precision by a rack and pinion or worm gear. It is a very common experience that every observer at the beginning of his observations adjusts the eyepiece

by moving it back and forth while seeing figures more or less out
of focus, until he finds the position in which the figure shows the
greatest clarity. The reason for this procedure was given just above.
But it is more important to point out the common and familiar fact
that while the eyepiece is being moved back and forth, the figure
seen does not alter its position in depth at all. In other words, the
observer sees changes in the clarity of the outlines and details of an
immovable figure. This happens, it should be noted, while the
centers of the waves emerging from the eyepiece and impinging
upon the eye are displaced by enormous distances, from a few
inches in front of the eye to infinity, and from infinity (in back of
the head) to a few inches behind the eye.

**187.** The same optical system, consisting essentially of a converging objective and a coaxial eyepiece, is called a microscope if
the dimensions of the focal lengths are interchanged, being small
for the objective and proportionately large for the eyepiece.

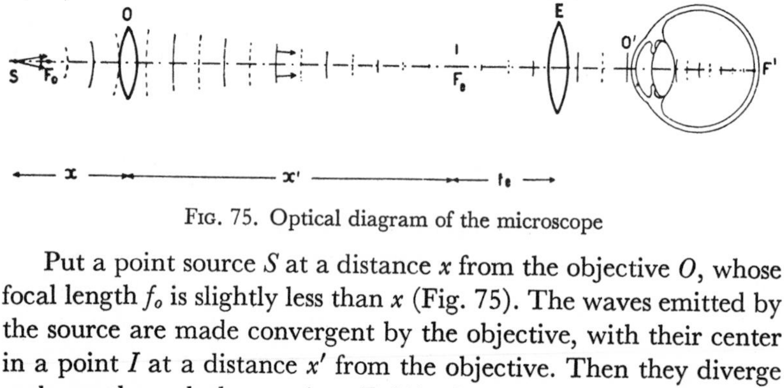

Fig. 75. Optical diagram of the microscope

Put a point source $S$ at a distance $x$ from the objective $O$, whose
focal length $f_o$ is slightly less than $x$ (Fig. 75). The waves emitted by
the source are made convergent by the objective, with their center
in a point $I$ at a distance $x'$ from the objective. Then they diverge
and pass through the eyepiece $E$. If its focus $F_e$ coincides with $I$, the
waves emerge plane. When they enter an eye $O'$, they are made
convergent with their center at $F'$. There the customary small group
of cones is stimulated. As a result a star effigy is created and placed
on the axis of the eye. The distance at which it is placed is controlled
exclusively by data drawn from the memory and the environment,
and does not depend at all on accommodation, the only optical
mechanism that could furnish any relevant information, if it were
taken into consideration.

That it is not taken into consideration is shown by the continual experience of all observers. In beginning a series of observations, they vary the focus of the instrument to find the conditions of greatest clarity. During these displacements (which are similar, if not exactly equal, to those discussed in § 186 for the telescope) the observer sees figures that are at times clearer and at other times less clear, but he never sees them change their position in depth.

The greater or lesser clarity of the figures depends on the curvature of the waves emerging from the eyepiece and on the possible compensation by accommodation. The fact that the figures remain motionless in depth during the process of focusing means that the effigies are always stationed at the same distance, whether accommodation intervenes or not.

In general, either because the observer usually knows where the object is, or because the instrument is ordinarily set up vertically, or nearly so, on a table which, as the observer is well aware, is not perforated beneath the instrument, the effigy is almost always placed in the plane of the object.

**188.** It is fairly easy now to account for the functioning of the microscope. If the source $S_1$ is not on the axis of the objective but is displaced from it by a segment $y$, the center $I_1$ of the waves emitted by the source and made convergent by the objective $O$ is at a distance $x'$ from $O$ and is displaced from the axis by a segment $y'$ such that

$$\frac{y'}{y} = \frac{x'}{x}$$

where $x$ is the distance from $S$ to $O$ (Fig. 76). As usual,

$$\frac{1}{x} + \frac{1}{x'} = \frac{1}{f_o}$$

The waves with their center at $I_1$ pass through the eyepiece $E$, which makes them plane and perpendicular to the direction $I_1E$. They enter the eye $O'$ and are rendered convergent with their center on the retina at a point $C'$ whose angular distance from $F'$ is $\gamma'$, equal to $y'/f_e$. The star effigy corresponding to the source $S_1$ is then placed in the direction $C'N'$ and in the plane of $S_1$.

The effigy's distance $y_e$ from the axis is obtained through multiplying $\gamma'$ by the distance from the object to the eye. This distance would be equal to $x + x' + f_e + d$, where $d$ is the distance from the eye to the eyepiece. But in these vague calculations it is useless to pretend to have very definite numbers. Hence all the refinements are dropped and we limit ourselves to approximations. Ignoring $d$ and $x$, which stand for very small distances in the arrangement

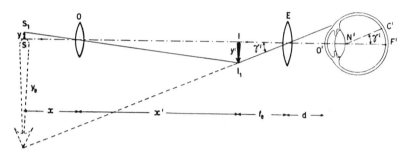

Fig. 76. Magnification of figures seen through a microscope

considered above, we regard $x'$ as constant and indicate it by $\Delta$. Hence

$$y_e = \gamma'(\Delta + f_e)$$

But $\gamma'$ is given by $y'/f_e$, and $y'$ by $y \dfrac{x'}{x}$, so that

$$\gamma' = y \frac{x'}{x f_e}$$

Substitute $f_o$ for $x$ here, and it follows that

$$\gamma' = y \frac{\Delta}{f_o f_e}$$

Hence

$$\frac{y_e}{y} = \frac{\Delta(\Delta + f_e)}{f_o f_e}$$

This equation tells us how many times $y_e$, the distance from the star effigy to the axis, exceeds the distance from the source $S_1$ to the axis (on the opposite side).

We may now repeat an argument advanced many times before. If there is a linear object $SS_1$, perpendicular to the axis, $y$ in length and at a distance $x$ from the objective, when the eye looks at it through a microscope, it gives rise to a linear effigy, $y_e$ in length, on the opposite side of the axis, and almost always located at the same distance from the eye as the object. Then the foregoing equation tells us how many times greater than the object this effigy is; in other words, it tells us how great the magnification of the microscope is, on the assumption that the effigy is placed in the plane of the object.

Both the telescope and the microscope can be constructed with a diverging eyepiece, put in front of $I$, the center of the waves made convergent by the objective. The analysis of this type of instrument would proceed, however, along lines so similar to those already followed in the case of converging eyepieces that the repetition is hardly worth the trouble. The only noteworthy difference is that the effigies created with a diverging eyepiece are oriented·like the object instead of being inverted.

**189.** It is convenient at this point to open a parenthesis for the purpose of glancing backward over our discussion. It may have been found a bit complex, while the conclusions may have been judged vague and not inevitable most of the time. Virtually every one of my readers, having already had occasion to peruse other treatments of "geometrical optics," so called, or "wave optics," so called, will now appreciate the simplicity, interrelation, and certainty of their conclusions, as well as the synthesis in mathematical equations that have become classical and generally accepted. Then a spontaneous question arises, Was it worth while to withdraw from so orderly a presentation for the sake of becoming involved in this other, which is so indefinite and complicated?

It is desirable to make a calm but unflinching comparison, to draw up a balance sheet of the merits and defects of both procedures. Recalling what was said in Chapter II, a critic may remark that this comparison was made three centuries ago. At that time the old philosophers who subordinated everything to the mind were replaced by the new philosophers who, profiting by Kepler's masterly and miraculous ideas, drove the ancient beliefs completely out of circulation, laid the foundations of the new optics, and thereon constructed that marvelous edifice which is now being subjected to scrutiny. Is this scrutiny really necessary?

The reply will be forthcoming at the end of the scrutiny. Too many students, not to say all of them, have heretofore unanimously answered this question in the negative. What is novel now is that I say, "Let us carry out this scrutiny." If seventeenth-century optics emerges victorious, the investigation will have been useless. But there are no incontrovertible reasons for saying that this will be the outcome. It may be the opposite.

**190.** Let us summarize the basic thinking of seventeenth-century optics. To avoid needless confusion, I shall always use wave language and translate into it the original language present in the terminology of geometrical optics.

The foundation on which seventeenth-century optics was built may be viewed as follows. From every point of a body waves are emitted in all directions. Consider only one of these points. Its waves are received by an eye and made convergent with their center on the retina. This point of the sensitive layer is stimulated. The stimulus is communicated to the mind, which represents it by a luminous colored star stationed outside the eye in the direction determined by the position of the affected point on the retina. The distance is measured by the eye, which is capable of feeling the curvature of the wave reaching the cornea. Therefore *the luminous point is located at the center of the wave arriving at the eye.*

This is the rule conceived by Kepler. He did not talk in terms of waves, as we have already observed, but he defined the famous "telemetric triangle." This triangle has its vertex at the material point under consideration in the object; its short side is the diameter of the pupil; and its two other sides are the straight lines joining the vertex with the ends of this diameter. To say that the eye feels this triangle, and thereby measures the distance from the vertex to the pupil, amounts to saying that the eye is in a position to feel the curvature of the wave reaching it. The long sides of this telemetric triangle are obviously nothing but the perpendiculars to the wave surfaces.

Hence direct vision was explained in the following way. By repeating the foregoing reasoning for every point of the object, the mind proceeded to reconstruct the figure of the object point by point. The observer could then be said to *see the object.* This was equivalent to saying that when an observer saw a figure, the corresponding object had to be there, provided no optical system came between the eye and the object.

When the waves emitted by a point $S$ are deviated by a plane mirror $MM$, they reach the eye deviated but not deformed. As is well known, their center is at a point $S'$, symmetrical to $S$ with respect to the plane $MM$. Therefore, on the basis of the foregoing rule the eye $O'$ must see the luminous point at $S'$. All eyes must do so always. Consequently it is possible to dispense with saying that the eye has to be present. The point $S'$ has its own position, determined independently of the observer. Hence it was called the *image* of the *object S*.

If instead of a point source $S$ there is an infinite number of sources constituting an object, for each of them the usual reasoning is repeated. The outcome is the reconstruction, point by point, of a figure that is symmetrical and symmetrically situated with respect to the plane $MM$ of the mirror. This figure is the image of the object, as given by the mirror. There is no longer any need for the eye to intervene.

**191.** Let us go on to the spherical mirror. If a point source $S$ is on the axis at distance $x$ from the mirror, the waves transmitted by the source to the mirror are reflected back by the mirror as either convergent or divergent waves with their center at a point $I$, whose distance from the mirror is a segment $x'$, connected with $x$ and with $r$, the mirror's radius of curvature, by the well-known relation

$$\frac{1}{x} + \frac{1}{x'} = \frac{2}{r}$$

This point $I$ is the center of curvature of the waves reflected by the mirror. Hence an eye which receives these waves *must* see the luminous point in $I$. It is, therefore, useless to reintroduce the eye into the discussion. Without further ado, $I$ is the image of $S$ as given by the spherical mirror.

At this juncture a small matter should be made clear. To go into it is almost pedantic, because so little is explained if it is clarified. Yet $I$, the center of the waves reflected by the mirror, is sometimes behind the mirror, at a point through which the waves never pass, as is true of a plane mirror. But at other times the center $I$ is in front of the mirror, at a point through which the waves actually pass. To distinguish between these two situations, the images of the first type are called *virtual*, and those of the second type, *real*.

If instead of a point source $S$ there is an infinite number of sources constituting an extended object, its image is built up by applying the foregoing construction to every point of the object. In the case of a concave mirror, virtual images are then found to be erect and larger, whereas real images are inverted and sometimes larger, sometimes smaller. If $y$ and $y'$ are the dimensions (in the direction perpendicular to the axis) of the object and the image, and $x$ and $x'$ are the corresponding distances from the mirror (that is, the distances connected by the equation given above) then the following relation holds:

$$\frac{y'}{y} = \frac{x'}{x}$$

On the other hand, if the mirror is convex, the images are always virtual, smaller, and erect.

**192.** If these are the rules of seventeenth-century optics, let us compare them with experience. Let us re-examine the cases considered above in §§ 190–191, where we were discussing vision by means of plane and spherical mirrors.

The plane mirror is brought forward as an immediately decisive proof of the perfect correspondence between experience and seventeenth-century optics. Its supporters, having shown that on the other side of plane mirrors symmetrical figures are seen symmetrically situated with respect to the reflecting surface, find the hypothesis demonstrated so conclusively that this is no longer a question of a hypothesis but of an incontrovertible *truth*. Then there is no longer any need of control experiments.

But the analysis in § 190 concerns mirrors placed a short distance from the observer. When the plane mirrors are far away, as in the experiment reported in § 128, the rule no longer holds. This experiment marks the first assault against the hypothesis of seventeenth-century optics in its innermost citadel, the plane mirror.

Let us go on to the concave mirror. With an object in the focus of the mirror, the reflected waves are plane (Figs. 43, 49). When the observer looks at the mirror, he should see virtual images infinitely distant; that is, he should see an endless hole behind the mirror, and at the back of this hole the image of the object should appear. *No one has ever seen anything of the sort.* In this connection seventeenth-century optics has been struck a most grievous blow.

Let us move the object out of the focus and closer to the mirror. The image should approach from infinity up to the mirror. Nobody has ever seen anything of the sort. For the image was not at first seen infinitely distant; and unless it was far away, it cannot approach. The hypothesis of seventeenth-century optics becomes more and more untenable.

Let us move the object out of the focus toward the mirror's center of curvature, as in Figs. 45 and 50. The image is real and behind the observer's head. The optics of the seventeenth century does not say what is supposed to be seen. If the observer is supposed to see the image, he would have to see behind his own head, and that has never happened. On the other hand, optics forbids any figure to be seen in front, because such a figure would have to consist of the centers of divergent waves reaching the eyes; and since there are no such waves, there should be no figure. The situation of seventeenth-century optics in this regard is most absurd. A like result follows when an attempt is made to apply the hypothesis under discussion to the other cases considered in Figs. 46 and 51.

In conclusion, seventeenth-century optics has shown itself to be utterly inadequate to explain the experimental data seen by anyone looking at a concave mirror. The fundamental hypothesis that "the eye sees a luminous point at the center of curvature of the waves reaching it" is almost never confirmed.

**193.** Let us turn to lenses. From our previous analysis the analogy between converging lenses and concave mirrors emerged so manifestly as to imply at once that the failure of seventeenth-century optics, as shown by the latter, must surely recur for the former. Nevertheless it is worthwhile making the test in detail, because the reader will probably be amazed by the results. He will be amazed by the power of the preconceptions that induced him to regard many demonstrations as true and to repeat them, although his experience constantly showed them proceeding in an entirely different way.

But let us get down to facts. Consider a converging lens with an object at its focus, as in Figs. 61 and 67. The waves emerging from the lens are plane. The observer should see virtual images behind the lens at an infinite distance. On millions of occasions observers (who have forsooth repeated this law and taught it) have looked at a sheet of paper on their desks through a converging lens, have even placed the lens so that the paper was in its focal plane, and *nobody ever* has seen a very deep hole in the desk with the figure of the paper at the bottom of the hole. Merely to say this makes us laugh. Yet this is what is required by the fundamental hypothesis of seventeenth-century optics.

Similarly if, while looking through the lens, the observer moves it closer to the paper, he should see the virtual image approach from infinity up to the lens. Nobody ever has seen anything of the sort.

If the position of the image of the paper is calculated by the equation for the distance from the center of the waves to the lens, it is found at once that the image should *always* be farther away from the lens than the paper is. Everybody knows that this *never* happens. The magnified image is seen in the plane of the paper.

Let us move the object away from the lens. The waves emerging from the lens must be convergent, with their center behind the eyes of the observer. What is he supposed to see? Nobody says. Yet no figures should be seen on the same side of the lens as the object, because on that side there never are any centers of waves emerging from the lens. Billions of users of converging lenses in their eyeglasses have shown for nearly seven centuries that things are exactly the opposite of what is required by seventeenth-century optics.

**194.** Let us go on to the diverging lens. The centers of the waves emerging from such a lens (when it is struck by plane or divergent spherical waves, like those emitted by material objects) always lie between the lens and its focus. For example, anyone looking through a $-4D$ lens should see the images of all the objects that are behind the lens in the region between the focus and the lens, that is, in an interval of 25 cm. Millions of experiences prove the contrary. All nearsighted people who wear diverging lenses in front of their eyes see the figures dispersed in space, as normal persons do. A normal person who puts a negative lens in front of one eye makes an effort of accommodation in order to see well (if he can do so) and then he sees as he would without the lens. A result more disastrous for seventeenth-century optics cannot be conceived.

**195.** To conclude this review a brief comment on the telescope and microscope will suffice. As we have seen (§§ 182, 185–187), when the eyepiece is moved closer to or farther from the objective, the emerging waves are divergent, plane, or convergent. If the famous fundamental hypothesis of seventeenth-century optics were confirmed, the observer should see the image nearby when the waves are divergent; he should see it at infinity when the waves are plane; and when the waves are convergent, who knows what he is supposed to see? As he moves the eyepiece along the axis of the objective, the observer should see the image approach and recede as far as infinity. Nobody has ever seen anything of the sort.

By now it is pointless to persevere in the proof. The basic hypothesis that "the eye sees the luminous point at the center of the waves reaching it" does not jibe with reality.

The doctrine that the virtual image is the figure seen, as sometimes happens in the case of plane mirrors, is one of the gravest errors ever committed in the study of optics.

*The image*, whether real or virtual, defined as the locus of the centers of the waves emerging from an optical system, *is a purely mathematical entity, entirely distinct from the figure seen*. Images have a definite position; the figures seen are created and located by the observer, and may be placed by one observer in one way, by another observer in another way. An image is a mathematical entity; the figure seen is a psychological entity. To have identified these two entities was a profound philosophical blunder. To convince millions of people that the two things are the same is one of the most ridiculous aspects of the teaching of science.

**196.** After this conclusion the reader will have to acknowledge, perhaps reluctantly, that the method pursued by us of taking into account not only physical factors but physiological and psychological as well has yielded good fruit. It has permitted us to describe the phenomenon of vision in a new way, which possesses no small advantage over the old way. For anybody who proceeds to make observations finds things going according to the rules and laws given here, whereas things almost never go as required by the principles heretofore considered established, which are based exclusively on physical elements.

After this comparison it is possible to appreciate to the full the great importance of Kepler's hypothesis of the telemetric triangle. Before it was put forward, experimenters used to observe directly in mirrors, prisms, and lenses. They found so vast a diversity of aspects that they obtained no conclusive result. The psychological intervention of the observer predominated to such an extent that it was impossible to arrive at any decisive factor of a physical nature.

By his masterly hypothesis Kepler eliminated the observer. As was remarked in § 46, he distinguished "pictures" from "images of things." The latter were what we have called the effigies created by the mind; the pictures were what was later termed "real images." Kepler said, "Forget about the images of things, deal only with the pictures," a valuable admonition that permitted the splendid organization and development of optics. But this was not the true optics, the science of vision. When the latter conception of optics was established, the eye and the observer had to be recalled to their stations. The images of things returned as the fundamental object of study.

But this change forced Kepler's telemetric triangle back to its proper function as a simple working hypothesis. It is in fact devoid of merit, because essentially, as we know nowadays, it amounts to saying that the observer locates the effigies on the basis of accommodation, a proposition that is not true at all. Today, however, after the enormous mass of research into the pictures, we can resume the study of the images of things with a much higher probability of securing laws of scientific and practical value than was possible three centuries ago, as the next Chapter will demonstrate.

But first of all it was necessary to get rid of the mistake made in the period when seventeenth-century optics was developing, the mistake that the images of things and the pictures were one and the same.

# CHAPTER V

# The Acuity of Vision

◇◇◇◇◇◇◇◇◇◇◇◇◇

**197.** As I have said and repeated often enough, the seventeenth century looked with great favor on Kepler's telemetric triangle, an idea that in our language may be translated into the rule that "the eye sees a luminous point at the center of the waves received by it." For this highly favorable attitude there were two reasons, which have already been indicated. The first and more immediate reason was the wonderful systematization conferred by that rule on optics, heretofore an extremely controversial and difficult subject in which no one had succeeded in establishing any order, however partial and provisional. The second reason was the support given to the rule by the philosophy of the time, that new empirical philosophy which was forging ahead of the old peripatetic philosophy, covered with glory but showing clear signs of exhaustion and inadequacy.

The rule of the telemetric triangle permitted optical problems to be solved by themselves without any need to have recourse to the eye, not to mention the mind. It permitted the construction of a *physical* science in the new sense of that word, namely, a science *independent of the observer*. True, the seeing was done by the observer (in those days nobody even thought of eliminating him) but his function was fixed; he located the effigies at the centers of the waves emerging from optical systems. These could, therefore, be examined independently of the observer, for it was sufficient to study the positions of the centers of the waves, that is, the images, whether virtual or real. Consequently the resulting optics may be called the "optics of images," or to be more explicit, the optics that investi-

205

gated the centers of the waves deformed or deviated by optical systems.

This method, which was none too clear and none too careful, produced a certain amount of floundering. The term "optics" had arisen because there were eyes; now an optics was being constructed without any eyes. To be sure, at first the intention was to construct an optics that would, so to speak, prepare the work of the eyes; but there was no thought that this could later lead to forgetting the eyes altogether, as actually happened. At first, talk about waves necessarily led to talk about eyes because, of the waves in the aether, only the visual were known, and of means of detecting them only the eye was known. Later on, invisible waves were discovered and other means of detection were found, such as photosensitive emulsions and photoelectric cells. Then there appeared nothing strange about lumping all that in the "optics of images," and something was constructed which, although called "optics," had absolutely nothing to do with the eye.

With these remarks I want to emphasize the conception put forward in Chapter I, in order to show that returning to "optics, the science of vision" is no mere question of words but a profoundly clarifying journey. Its purpose is to restore the rule of the telemetric triangle to its true function as a working hypothesis. However valuable and meritorious, the rule is nothing more than a working hypothesis. After the resounding success it gained in its true function, should it pretend to advance to loftier heights, it would be doomed to certain failure.

**198.** I have resumed this line of thought because now our investigation will go deeper and show how productive of error it was to have forgotten that optics is the science of vision and to have developed it blindly (for that is the appropriate word) as the optics of images or, I reiterate, as the optics of the positions of the centers of the waves.

Here too, as in Chapter IV, we shall examine a group of phenomena while keeping the eye in its place. We shall then recall the conclusions reached by the optics of images and compare those conclusions with experience.

Consider once more a point source of waves. We must now proceed with increasing caution. A point source cannot really exist, because a geometrical point has no dimensions and therefore cannot be material and emit waves. Hence, whenever a material source is called a point, it is assigned an attribute that certainly does not correspond with reality. It is possible to talk about point sources on paper, but not in practice.

Nevertheless there are sources that *seem* to be points, for example, the stars. They are an interesting example because everybody knows that although they *seem* to be minute points, they certainly are not. On the contrary, in order to be visible at such vast distances they must be so big that the sun is modest in comparison.

If they seem to be points, then they seem so to somebody, that is, to an observer, understood in the broadest sense of the term as not only a person but also a device. In other words, the description of a source as a point is not absolute and does not refer to a trait of the source itself (for if it were understood in this geometrical sense, no material source, I repeat, would be a point) but it depends on the means by which the source is observed.

A source is said to be a point when a given observer fails to perceive that it is not. The same source, as is at once obvious, may be a point for one observer and not for another. This remark will appear a mere commonplace to anyone who considers that an object so small as to be called by everybody a barely perceptible point when observed with the naked eye becomes a whole world if examined with a microscope. What is a point for the naked eye may no longer be such for an eye equipped with a microscope.

**199.** One of the fundamental requisites of optics is to determine the dimensions of a "source that seems to be a point." This expression, being a bit too long and not very attractive, has been replaced by another term, namely, "optical point source." But this too is rather long and is, therefore, enunciated in full only when its content must be made clear. Otherwise, simply "point source" is used with the understanding, however, that it is said in relation to the conditions of observation. Those who wish to talk in pure abstractions may consider a geometrical point source also.

The problem that I propose to take up is to determine the dimensions of an optical point source. Merely from the remarks made in § 198 (to which no objections can be raised seriously, so far as I am aware) it follows at once that in order to determine those dimensions, the observer must be defined. Our task naturally includes the naked eye in direct vision as well as the eye equipped with optical systems.

**200.** Let us begin with the naked eye, and perform the following experiment. Put an observer several dozen yards away from a surface that emits visual waves and can be diaphragmed with progressively smaller diaphragms. Every time the diaphragm is changed, ask the observer whether he perceives the dimensions of the surface or not. A certain diaphragm will have to be reached below which the dimensions are no longer perceived.

But an experiment of this kind would not lead to definite results. A new factor intervenes that heretofore has been mentioned only vaguely but that will hereafter acquire ever greater importance, namely, the energy of the waves received by the eye. The defects of the eye intervene too. They are never absent, and they vary a good deal from one eye to another, even in the same person. They produce the complex phenomenon known as *irradiation*, which was described in § 106.

Returning to the diagram in Fig. 24, let us resume the discussion initiated in § 84. The waves emitted by an ideal point source $S$ enter the eye and are made convergent with their center on the retina at the point $F$ in the fovea. The center of the waves, if they are spherical as has been supposed, is a point. Not entire hemispheres are involved, however, but only segments of waves. Hence we know that near the center the energy is not concentrated in a point (this would make no sense, physically speaking, because a point has no dimensions, and therefore to be concentrated in a geometrical point means not to exist) but is distributed in a centric (§ 328). Consequently what is affected on the retina is a whole area, which in § 89 we estimated as perhaps four or five cones. As was indicated in § 88, there is no need of a more accurate determination based on the known phenomena of diffraction. That can be done when the waves are truly spherical. Although after passing through the cornea, crystalline lens, and lacrimal, aqueous, and vitreous humors the convergent waves emerge in an average form close to the spherical, nevertheless they are in general affected by certain subtle ripples and irregularities that perceptibly alter their concentration on the retina.

As was remarked in § 106, when the eye is turned toward a point source, the distribution of energy on the retina shows a maximum at the center with many offshoots. If this distribution could be represented by a solid, there would be a central mass with various diminishing spurs, some longer, others shorter, distributed irregularly around the central peak.

As a result the brain receives impulses from the cones affected by not only the central stimulus but also the surrounding stimuli. The effigy created, accordingly, comes to have the form of a "star," that is, a central nucleus surrounded by many bristles radiating outward, some longer, others shorter, called the *rays of the star*.

**201.** What are the dimensions of these radiated stars? This is a highly complex question, because these dimensions are a function of many factors. We must bear in mind, first of all, that the sensitivity of the retina is not infinite, and therefore stimuli below a certain limit are not perceived. In other words, a *lower threshold* exists; when a stimulus fails to reach this threshold, it is not felt. This concept of a lower threshold is implied by the "*all-or-none law*," which we used in § 85 to explain the functioning of the eye. For if the energy does not attain the minimum quantity required to discharge at least one impulse, no information arrives at the brain.

We must further take into account that the retina is a slightly diffusing translucent membrane. Before reaching the rods and cones (which have their free ends turned toward the external coatings of the eyeball) the radiation must pass through the whole layer of ganglions and nerve fibers running from every sensitive element to the junction of the optic nerve with the retina. Hence when energy in the form of a stimulus impinges upon this membrane, it is diffused somewhat in the zone surrounding the one struck. An effect of the same kind results from diffusion in the humors which, especially in elderly people, show a not inconsiderable turbidity. Also the crystalline lens with its cellular structure, as well as the cornea and the epithelial tissues, cause a perceptible diffusion of the radiation passing through them.

It should be recalled too that the diameter of the pupil varies from a minimum of 2 mm to a maximum of 8 mm. The variation is generally regulated involuntarily as a function of the intensity of the energy flux incident upon the retina.

Finally it should be pointed out that the effect of the radiation on the retinal substances is not always the same. Not only is it a function of the wave length, but it also varies with the structure of the photosensitive retinal substance. Thus there is the so-called *adaptation* of the retina to darkness; that is, changes in the retina occur on account of which the chemical or electrical effects due to the incident radiation are more or less conspicuous. These are phenomena about which little is known as yet. But if we want a rough sketch of them, we may compare them to the differing speeds of photographic emulsions. The retina has the power to modify the structure of the photosensitive substance so that its yield of nerve impulses is highest when the incident radiation is weakest, and on the other hand becomes moderate when the radiation is strong.

**202.** Now consider a very minute source of visual waves, such as a star in the sky; for the stars are certainly points, so far as the naked eye is concerned. The waves reach the eye and impinge upon the retina. The effect, as is obvious at once, is a function of the intensity of the source.

If the source is very weak, so that the retina is barely affected by the central peak, the spurious offshoots fail to reach the retina's threshold of sensitivity. Hence, for the purposes of vision, it is as though they did not exist. The same holds true for the radiation diffused in either the retina or the humors, for its effect is nil. On the other hand, if the source is the only thing in front of the eye, the pupil is dilated to the maximum, and the retina is adapted to maximum darkness; that is, the photosensitive retinal substance shows the maximum sensitivity. Under these circumstances the number of affected cones is reduced to the minimum, and the effigy created is a luminous spot of inappreciable dimensions. However, to achieve these conditions, it is necessary to have absolute darkness, a very feeble star, an interval of time long enough to attain the required adaptation, and an observer specially trained to place the center of the waves in the center of the fovea.

For the sensitivity of the retina is not uniform throughout its entire surface. The minimum is found in the fovea, and the maximum in the so-called *parafoveal region*, a ring around the fovea. When the radiation on arrival is so weak as to require the maximum retinal yield in order to be perceived, the eye automatically rotates a little for the purpose of putting the center of the waves in the parafoveal region.

But here other complications intervene. Now there are no longer only cones involved, but also rods. Moreover, the connections with the nerve fibers are already in clusters; that is, various retinal elements are linked with the same nerve fiber. But the lower threshold of sensitivity is different for rods and cones. The latter seem to be much less sensitive than the former, so that diurnal or *photopic* vision is said to be performed with the cones, and nocturnal or twilight or *scotopic* vision with the rods.

As is evident, the picture is far from simple, and the features that influence the result are still not exhausted. As our discussion of the situation continues, the reader will become increasingly aware of how much caution is necessary before reaching a conclusion having to do with the eyes.

Our first result, then, is that when the source's intensity is such that it is close to the retina's lower threshold of sensitivity, and the source's dimensions are such that the number of stimulated retinal elements is the minimum (not precisely defined), the effigy created is a "point," in the sense that the mind associated with the affected eye never creates a smaller effigy. However, these conditions are so unusual that they do not deserve further consideration.

**203.** Now suppose the intensity of the star increases slowly. In addition to the central peak, the most important surrounding offshoots start to make their presence felt on the retina. But they do so only in the part near the peak, because they weaken as they extend outward and soon fall below the threshold. The effigy created is brighter than the previous one, and commences to show some tips along its sides.

As the source becomes more intense, increasingly longer segments of the offshoots cross the threshold. Some of the offshoots that had not reached the threshold before begin to affect the retina now. Hence the effigy exhibits a constantly growing number of rays, of which the most intense grow steadily longer. Under these conditions rays may be so long that the star occupies a cone several degrees in aperture.

Hold the intensity of the source at a fixed value. But instead of keeping the source in an absolutely dark field, have the background emit a diffuse radiation. If this is very weak, the effigy of the star does not change. But if the radiation from the background becomes noticeable, the effigy of the star progressively loses its weakest rays and the most intense rays shorten.

These effects are due in large measure to the fact that under the action of the radiation coming from the background, the pupil begins to contract. As a result the retinal zones affected by the radiation of the star shrink too. For when the pupil contracts, the first result is a decrease in the flux of energy, which obviously is proportional to the area of the pupil. Therefore, the weaker offshoots and the tips of the more intense fall back below the threshold of sensitivity. In addition, when the optical system of the eye is diaphragmed, the irregularities that act on the waves are diminished too, thereby altering the distribution of the energy around the center of the waves. Hence the resulting effigy comes out adorned with a smaller number of rays, while those present are shorter.

The case now being described may be clarified by a familiar observation. Let us recall the effect that many people notice in looking up at a clear sky at night. When the moon is not visible, many stars are seen surrounded by rays, which are longer and more plentiful in the most brilliant stars. On nights when the moon is full, the visible stars are far less numerous and have few if any rays.

**204.** Let us resume increasing the intensity of the source. Opposing phenomena occur. On the one hand, the pupil contracts and the sensitivity of the retina drops. On the other hand, those offshoots on the retina that surpass the threshold lengthen; weaker offshoots begin to be perceived; diffusion in the substance of the retina and humors of the eye becomes increasingly noticeable. The first two changes tend to reduce the affected retinal zone, whereas the others tend to enlarge it.

Within a certain range of the source's intensity the two opposing sets of actions balance each other. The eye, so to speak, succeeds in confining the zone within narrow boundaries. But finally, when the pupil has reached its lower limit and the sensitivity of the retina has done the same, there are no further compensatory actions. As the intensity of the source mounts, the affected retinal zone also grows considerably. The effigy created is an expanded, brilliant, blinding star, adorned with very long rays that occupy a great part of the field of vision, while the background too appears somewhat luminous.

Everyone has had occasion to see figures of this kind when his eyes have been directly illuminated at night by the headlights of an automobile, even when it was rather far away. The sense of discomfort felt by the observer under these circumstances is well known. We say that he is *dazzled*. The phenomenon is evidently very complex, and if we wished to give an exhaustive presentation of it, we should need a series of curves. A personal element would always intervene, however, such as the extent and distribution of the irregularities in the eye making the observations.

To have reliable data, it would be necessary to derive an average from measurements made with a large number of eyes. So far as I am aware, such measurements have not yet been made. Without entering into so complicated a subject, I confine myself to the remark that things go badly when the source is either too weak or too strong. There is an intermediate range of intensity, as we may call it, within which things go better. In other words, vision of a point source is a phenomenon which, considered as a function of the intensity of the source, gives optimum results under intermediate conditions.

**205.** Of considerable interest for what we shall have to discuss later on is the observation of a point source through an *artificial pupil* or hole, less than 2 mm in diameter, in an opaque plate. As a result the aperture of the waves entering the eye is limited, not by the natural pupil, but by the rim of the hole. As we already know from § 63, the diameter of this hole is a very important factor, usually indicated by $D$.

When $D$ drops below 1 mm, in almost all eyes the irregularities of the optical system do not make themselves felt, even if the point source is very intense. This means that the effect is due not so much to the decrease in energy as to the fact that the central zone of the crystalline lens ordinarily is the best from the structural point of view. When this is not the case, the eye is very defective.

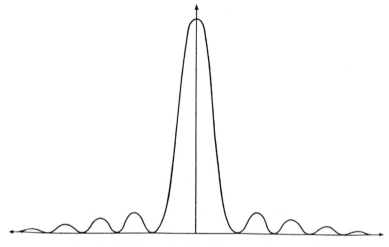

Fig. 77. Distribution curve of the energy in a centric

Then if the irregularities do not make themselves felt, the waves propagated in the eye behind the crystalline lens are sensibly spherical, and around their center on the retina the energy is distributed in a centric (§ 326). Fig. 77 shows the structure of this distribution, which has been studied with great care. There is a maximum in the center, with flanks diminishing to a minimum in the form of a ring, which reaches the value of zero along a circle whose radius in general is

$$r = \frac{1.22\lambda}{D} f$$

In our case

$$r = \frac{24.4\lambda}{D} \text{ mm}$$

since the focal length of the eye may be considered to be 20 mm.

Then follows a ring with a secondary maximum, much inferior to the central maximum; another ring of zero intensity; another ring with another secondary maximum, slightly inferior to the preceding maximum; and so on.

When this distribution of energy reaches the retina, the zones in which the effects may be felt differ according to the intensity of the source.

**206.** Let us begin with a very weak source, one so weak that only the central maximum surpasses the threshold of sensitivity. The number of affected cones (on the assumption that the observer succeeds in keeping the centric in the center of the fovea) is the smallest possible. This number depends to a slight extent on $D$, the diameter of the hole, because the maximum in the curve of the centric is flattened out somewhat. Consequently the effigy created is a tiny disk. As the intensity of the source increases, the brightness of the disk grows, but its diameter expands too.

When the intensity reaches a value high enough for the first secondary maximum also to cross the threshold of sensitivity, a luminous ring, very faint and very thin, begins to be seen around the central disk. Then, as the intensity of the source continues to mount, the ring keeps the same radius but gains in thickness; that is, the dark ring separating it from the central disk narrows a little, while the central disk itself continues to expand. Meanwhile a second ring appears, very thin at first and later thickening constantly. Then a third ring follows, and so on. With very intense sources it is possible to see as many as five or six rings. In the meantime, however, the first dark ring has become extremely thin.

The effigy thus created is located out in space, in the direction of the source of waves, on the basis of data derived exclusively from the memory and environment, if not entirely from the imagination. For even if the mind wanted to use the information furnished by accommodation, in this case such information is highly uncertain on account of the effect of the artificial pupil, the information being more indefinite the smaller the diameter of the hole.

**207.** If the source emits pure waves, the mind endows the effigy with a certain spectral color tone, based on the modulated frequency of the impulses reaching the brain. When the source is so weak, however, that the central maximum barely succeeds in crossing the sensitivity threshold of the retinal region affected, the frequency of the impulses is so low that we can no longer talk about carrier frequency and modulated frequency. Then the mind can no longer assign a definite hue, and a gray spot is generally seen.

Fig. 78. Centric

On the other hand, when the source emits all the visual waves, the individual centrics corresponding to the various waves have different dimensions. The disks corresponding to the longest waves have almost double the diameter of those corresponding to the shortest waves. The resulting effect is translated into a coloration of the rims of the disk and rings. The rim of the disk is red; so is the external rim of the rings, while their internal rim is blue. Apart from the coloration, the aspect of the effigy seen under these conditions is shown in Fig. 78.

The smaller the hole, the larger the diameter of the effigy, if we disregard the effects of the intensity. Thus if we observe in succession a point source of intensity $k$ through a hole 0.5 mm in diameter, and a source of intensity $4\,k$ through a hole 0.25 mm in diameter, the flux reaching the retina is identical. But in the latter case it is spread out over an area four times greater, and therefore the diameter of the effigy seen is about double the other.

**208.** Let us keep our attention fixed on this type of phenomenon. Suppose we observe a point source through a hole of constant diameter, so small that the distribution of energy on the retina is not disturbed by the irregularities of the eye's optical system. The effigy created by the mind will, therefore, have the aspect of Fig. 78, with well-defined dimensions when the mind has decided to place it at a given distance.

Now consider another point source, equal to the former and situated at the same distance from the eye. An energy distribution similar to the previous is formed on the retina. The mind accordingly fashions two equal effigies, placed at the same distance in depth and in directions determined by the positions of the stimuli on the retina (Fig. 79). The two effigies will generally be distinct from each other, because the stimuli on the retina will be distinct.

Now let one source approach the other. The affected retinal zones approach each other too, and begin to overlap. As the two sources continue to draw closer together, the

FIG. 79. Two centrics at varying distances

zones overlap more and more, until finally the superposition becomes so complete that it is impossible to say whether the effect is due to two separate point sources or to a single one of double intensity.

This is a highly interesting phenomenon, which should be examined somewhat in detail. Suppose the distance on the retina between $C_1$, the center of the first centric, and $C_2$, the center of the second, is a small fraction (say, 1/10) of $r$, the radius of the first dark ring of each centric. Or rather, considering the slight contribution made by the rings of centrics to the final outcome, we shall ignore the rings altogether for the sake of simplicity. Every centric is thereby reduced to a simple disk, in which the distribution of energy along a diameter is shown in Fig. 80$a$. On the ordinate, $E$ indicates the *irradiance* (§ 78) or flux of energy impinging upon a unit of the retinal surface. $C_1$, the origin of the coordinates, corresponds to the center of the first centric. Merely for the purpose of having a standard of reference, this center may be imagined to coincide with the center of the fovea.

**209.** If we consider also the second centric with its center at $C_2$, we have another energy distribution equal to the former, but displaced by $r/10$. Irradiances may be added arithmetically at every point (if the two sources of waves are not coherent). For that

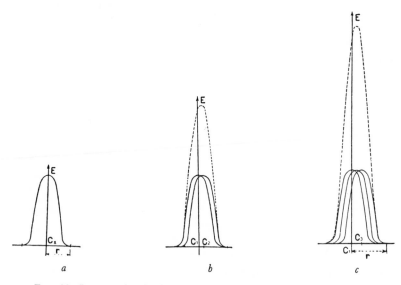

Fig. 80. Increase in the intensity of irradiance on the retina, when the limits of an

reason the simultaneous presence of two centrics on the retina gives a combined energy distribution like that indicated by the broken line in Fig. 80$b$. The combination differs imperceptibly from a centric; its center of symmetry is displaced by $r/20$ from $C_1$ and $C_2$; its ordinates are sensibly double those of Fig. 80$a$; and its abscissas have undergone a very moderate increase of scarcely 5 per cent.

The resulting effigy is therefore a slightly oval figure (the increase of 5 per cent occurs in only one direction) whose brightness is greater than that of the effigy created under the stimulus of a single centric. Some investigators maintain that the subjective

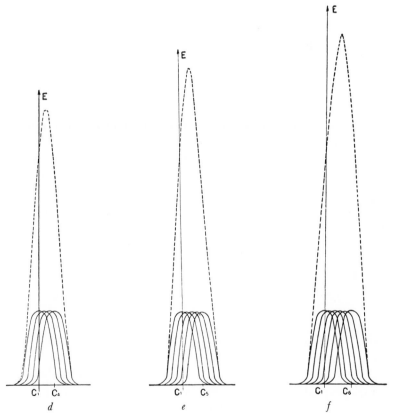

the dimensions of the source are increased, while remaining within optical point source

brightness of the effigy varies with the logarithm of the intensity of the stimulus or irradiance. But this rule is not firmly established. Besides, the value of this brightness is of no interest to our discussion. Consequently I shall limit myself to noting its increases and decreases without specifying the amounts.

Given the indefiniteness of the rims, ovalness of 5 per cent may pass unnoticed. Even if this were not so, a similar analysis could be repeated by putting the distance between the centers $C_1$ and $C_2$ equal to $r/20$, $r/40$, . . . ; certainly there must be a denominator big enough to make the ovalness unobserved. Under those conditions (and for all smaller values of $C_1C_2$) the only effect of two simultaneous centrics is to double the irradiance on the retina and accordingly to increase the brightness of the effigy, since the form of the combined figure is sensibly equal to that of each component. Hence for all values of $C_1C_2$ less than this limit *the effect of the superposition is exclusively photometric.*

Because the essential content of our reasoning does not change with a variation in the limiting value of $C_1C_2$, let us assume it to be $r/10$. Besides, we should note that no one value is applicable to all cases, for two reasons. In the first place, it is a disk with indefinite rims whose ovalness is to be estimated, a process strongly influenced by the observer's skill and training. Secondly, as was pointed out in § 206, the apparent dimensions of the disk in the effigy depend on the intensity of the source, whereas $r$ is a quantity independent of this factor. Therefore $r/10$, taken as an example, is not outside the range of possible values.

**210.** Now let there be a third point source, equal to the other two and aligned with them, the third being exactly twice as far from the first as from the second. On the retina there will be a third centric, with its center $C_3$ aligned with $C_1$ and $C_2$ at a distance $2r/10$ from $C_1$. The effigy created is a figure like the previous one, but it has an ovalness of 10 per cent and greater brightness (Fig. 80c). Add a fourth source, then a fifth, all aligned and spaced at equal intervals. The figure seen increases in ovalness (reaching 25 per cent with five sources) and its brightness is considerably enhanced; in fact the irradiance on the retina is augmented almost five times, as shown in Figs. 80d and 80e.

However, if we continue to add equal point sources, aligned and spaced at equal intervals, the photometric increase henceforth takes place in progressively smaller steps (Fig. 80*f*). With ten sources we reach the maximum irradiance on the retina, and there-

fore a brightness of the effigy approximately equal to that attained with nine, because when the sum of the ordinates is added up, the contribution of the centric with its center at $C_1$ is very minor.

If we continue to add as many aligned point sources as we like, the effigy grows proportionately longer but its brightness stops increasing. Indeed the sum of the ordinates at every point, except those at a distance less than $r$ from $C_1$, consists of ten equal magnitudes (Fig. 81).

**211.** The foregoing discussion contains the criterion that permits us to describe a source either as a point or as a linear object. The distinction may be made clear by the following procedure. Begin with an ideal point source, and add one after another in a line so as to make a linear source, as long as required. At first the observer sees a very weak centric. Then, while continuing to watch it, he notices that it grows brighter and that its diameter at the same time enlarges slightly, on account of the increase in energy, as was

FIG. 81. Transition from a point source to an extended source

mentioned in § 206. But its form does not change. Later the *increase* in brightness becomes less conspicuous, while the figure begins to look oval. Finally the brightness ceases to be augmented; that is, it reaches its upper limit. The figure definitely lengthens out and, taking the shape of a wire, continues to grow longer, changing its form without altering its brightness, so long as the number of sources continues to rise. In this way the effigy, which initially had the characteristic form corresponding to a point source, in the end assumes the characteristic form corresponding to a linear source.

Now let us proceed in the opposite direction. Put in front of the eye a linear source and reduce its length progressively. At first the corresponding effigy is an elongated figure of a certain width and brightness. Later it begins to shorten, while keeping its width and brightness unchanged. When the length is about twice the width, its decrease becomes slower, while the brightness begins tc decline slightly. Then the decrease in length becomes barely perceptible, while the brightness diminishes rapidly. The result is a figure whose ovalness is no longer appreciable. But as the source continues to be shortened, a drop in brightness is observed, together with the consequent contraction of the diameter. Then if the threshold of retinal sensitivity is reached, and the source is shortened further, nothing is seen any more.

**212.** In this last phase, when further reduction of the source produces a decrease in brightness but no change in form, the source clearly behaves exactly like an ideal point source of variable intensity. It should then be said to be *optically a point for that observer*. His mind in fact has no optical information which reveals the linear form of the source to him.

In the initial phase, on the other hand, when the reduction of the source produces a shortening of the effigy without any alteration in its brightness and width, the source should be called definitely *linear*.

In addition there is the phase of transition from one typical form to the other. In this phase diminution of the source is accompanied by a reduction in the length (or rather in the ovalness) of the effigy and at the same time in its brightness. For this kind of source there is no designation, at least for the present. The general practice is to assign half of this phase to the point source and half to the linear source; in other words, the source is called a point even if the effigy is slightly oval.

Apart from this transitional phase, we may now speak of an optical point source, not in any ideal, mathematical, and physically absurd sense, but in a sense that can be realized in practice. Some

numerical data may be not without use in serving to clarify the
foregoing concepts.

**213.** Suppose we put in front of the eye a diaphragm of diameter $D = 0.5$ mm. The angle $\gamma$ which $r$, the radius of the centric's
disk, subtends at $N'$, the nodal point of the eye, is given by

$$\gamma = \frac{1.22\lambda}{D} = 2.44\lambda$$

When a source's extreme points give rise to centrics whose centers
are separated by a distance exactly equal to $r$, we may take that
source as the halfway station, in the sense explained in § 212, between a point source and a linear source. Such a source should be
called a point when its length subtends the angle $\gamma$ at the eye.
If $x$ is its distance from the eye, its linear dimension $y$ is given by

$$y \leqq 2.44\lambda x$$

If we put $\lambda = 6(10^{-4})$ mm, we have

$$y \leqq 14.64x(10^{-4}) \text{ mm}$$

For example, at 5 m from the eye,

$$x = 5000 \text{ mm} \quad \text{and} \quad y \leqq 14.64 \times 5000 \times 10^{-4} \leqq 7.32 \text{ mm}$$

Any filament that emits visual waves having a wave length of
$6 \times 10^{-4}$ mm and that is shorter than 7 mm behaves like an optical
point if it is at least 5 m from an eye diaphragmed with an artificial
pupil 0.5 mm in diameter.

If the diameter of the artificial pupil were reduced to 0.25 mm,
the length of the optical point source would obviously be 14 mm at
a distance of 5 m from the eye. If the diaphragm were contracted
still further, the dimensions of the optical point source would increase in inverse proportion.

**214.** The discussion of a source linear in one direction may be
repeated for any other direction. This process leads to the definition
of an *extended* source as being, for a given observer, a source having
dimensions larger than a disk whose diameter corresponds to that
of the greatest point source, for that observer.

It has been possible to establish these conditions rather easily for an eye equipped with an artificial pupil small enough for the effigy to have the aspect of a perfect centric. When the pupil is bigger and irradiation occurs, the analysis becomes far more complicated. Personal elements intervene, because every eye has its own irradiation. The intensity of the source affects the result much more obviously, since changes in this intensity cause the irradiation to vary much more than the dimensions of a centric's disk. Moreover, the effigy produced by a point source in the presence of irradiation lacks symmetry of rotation, and therefore equal effects are not felt in all directions on the retina. Nevertheless the criterion for determining the dimensions of a point source for an eye with an artificial pupil may be applied to the naked eye also, even if the numerical results differ from one eye to another.

**215.** To conclude our discussion of the point source I wish to emphasize that the minimum dimensions of a source are determined, not by geometrical factors (that is, by its form) but by energy factors. In other words, a source is too small to be seen when the waves that it projects upon the eye are incapable of stimulating the retina or making impulses reach the brain. Only then, when it receives no information, does the mind fail to create any effigies.

If the energy conveyed by the waves to the eye is enough to cross the threshold of sensitivity, then an effigy is created according to the rules set forth above. In other words, so long as the source is kept below the dimensions of the maximum point source, the effigy has an aspect corresponding, not to the shape of the source, but to the defects of the eye and the effects of diffraction. Only when the dimensions of the source surpass this limit does information about the shape of the source begin to flow to the mind, and only then does the corresponding effigy commence to represent that shape.

On account of the importance of this topic it is desirable to trace the changes in vision as the dimensions of the source vary continuously. For example, consider the global portion of a frosted electric bulb that is lit. Every point of the globe emits waves that enter an eye directly. On its retina an area with a circular boundary is affected. By means of the impulses sent to the brain by way of the optic nerve with a certain frequency, the mind is led to create an effigy in the form of a disk with a circular rim. This effigy will have a certain brightness and will be located, say, exactly on the bulb.

Now, along the path by which the waves reach the eye, move the bulb away from the eye. Since the angular magnitude of the globe diminishes, the retinal zone affected decreases too; so does the effigy, but its brightness does not.

This is true until the globe reaches the dimensions of the point source for that eye. From then on the effigy has the form, not of a disk, but of a star surrounded by the rays characteristic of that eye. As the source draws farther away, the rays become shorter and less numerous because the weakest disappear. The intensity of the star drops continuously until the effigy contracts to a point, which grows steadily weaker until finally it is extinguished.

**216.** For the purpose of further accustoming the reader to this way of viewing the subject, let us observe a lighted incandescent lamp, if possible of the type having a nonspiral tungsten filament. Looking at such a wire is rather unpleasant, because the waves that it projects on the retina carry too much energy. Naturally the pupil contracts to its minimum size, and the retina too reduces its sensitivity as much as it can. Nevertheless the effects of irradiation are still conspicuous. If the lamp is 30 cm from the eye, however good this may be, it sees in place of the filament a ribbon about two mm wide. The width of the filament is just a few hundredths of a mm. Obviously the brilliant ribbon seen by the observer is an effigy created by his mind on the basis of information reaching it from the retina, where the stimulated zone is so wide on account of irradiation.

Now remove the lamp to a distance of 60 cm from the eye. The brilliant ribbon becomes double what it was before, whereas if it were an effect of the material dimensions of the wire, it should become half. Instead it becomes twice as wide, while its length, which corresponds closely to that of the wire, becomes almost exactly half. For the width of the affected retinal zone depends, not on the size of the source, but on the irradiation or the defects of the eye, and therefore remains constant. Since its angular magnitude is constant, and the effigy is located twice as far away as before, the width of the effigy must double.

If we want to reduce this width while keeping the distance from the lamp to the eye unchanged, we must diminish the effects of the irradiation. This may be done in two ways, on the basis of what was said in §§ 205–206. Either contract the diameter of the pupil with a diaphragm whose aperture has a diameter less than 2 mm, or weaken the irradiance on the retina, for example, by means of an absorbing glass. If we look at the filament through a glass whose absorption increases, we readily observe the width of the luminous ribbon becoming steadily narrower, and when the brightness is decreased to a very low level, the effigy too is cut down to a very thin wire.

**217.** When there are two point sources (and in talking about them we now know that the term "point source" is no longer devoid of practical significance) the situation is so important that it deserves to be examined with great care.

In seeking to determine the dimensions of an optical point source, we began in § 208 by observing what occurred when two point sources in front of the eye approached each other. Then in § 209 we dealt with the extreme case, in which the effigy provoked by the action of both sources on the eye was sensibly equal (apart from its brightness) to that provoked by a single source. In §§ 210–212 we considered a very large number of sources in a line, for the sake of finding the demarcation between a point source and a linear source for that observer.

Now we must return to the analysis of what happens when two point sources approach each other. But our present purpose is to discover how vision of two separate figures passes over into vision of a single figure. Again it is proper to advance by stages, starting with the simplest and most routine case. In this we put in front of the eye an artificial pupil with so small a diameter that the effigy corresponding to a point source is a typical centric. Then the sequence of the appearances that result when two sources approach each other is once more that of Fig. 79.

When the observer is in a position to decide that there are two sources, he is said to separate or *resolve* them. Our task, then, is to determine the limit of such resolution. The arguments that lead us to fix the limit of resolution follow the model of those that led to establishing the limit of the point source. If the two sources are so little distant from each other that, taken together, they fall within the maximum dimension of the point source, then clearly they cannot be resolved under those conditions by that eye. Hence we conclude at once that the limit of resolution is a function of the traits of the observing eye as well as of the intensity of the source.

**218.** Reviewing the considerations by which we established the limit between a point source and a linear source for a given observer, we may say that this limit differs from the limit of resolution in the following way. In the former we notice a lengthening or rather marked ovalness of the figure, whereas in the latter we observe two maxima of brightness separated by a darker strip.

In other words, before resolution can occur, the mind must be informed that the sources are two rather than one, however elongated. For information of this sort to reach the mind, the stimulus on the retina must not be of equal intensity along its entire major dimension. On the contrary, its central portion must show a decrease *appreciable* by the sense mechanism, and then the two maxima are perceived as separate.

When the problem is posed in these terms, the intensity of the source is clearly of fundamental importance. For, together with the sensitivity of the retinal layer, it determines the dimensions of the affected retinal zone, as we learned in §§ 201–204.

Put the two sources in a given place. Then $C_1$ and $C_2$, the centers of the disks of the corresponding centrics, have definite positions on the retina, and therefore the distance $C_1C_2$ is fixed. If the disks of the centrics are so small that their radius $r$ is half, or less than half, of the distance $C_1C_2$, there is certainly a minimum of irradiance in the middle. But if the radius $r$ is equal to or greater than $C_1C_2$, what happens in the middle is uncertain. There may even be a minimum, but one so little different from the two maxima in $C_1$ and $C_2$ that the difference escapes detection by the sensitivity of the retina.

In §§ 206–207 we saw that the radius of a centric's disk depends on the intensity of the source, the diameter $D$ of the artificial pupil of the eye, the wave length $\lambda$ of the radiation emitted by the source, and the retina's threshold of sensitivity, adaptation being taken into account. To complete the picture of the principal elements that influence this phenomenon, we must ascertain the sensitivity of the retina to differences of irradiance. The answer has been given experimentally, as shown in Fig. 82. The ordinate represents the reciprocal of *Fechner's fraction* $\Delta E/E$, the least perceptible difference. The abscissa indicates the logarithm of the irradiance $E$. Meanwhile I wish to emphasize that here too the predominant influence

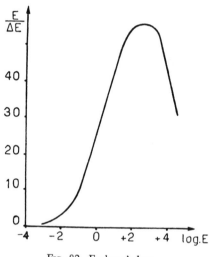

FIG. 82. Fechner's law

is the irradiance. For Fechner's fraction increases as the irradiance weakens or when it becomes too strong.

**219.** Now we can summarize and synthesize the composite effect of all these factors.

In a general way we may say that the limit of resolution depends, in inverse ratio, on the diameter of the artificial pupil (provided that this leads to the creation of an effigy with rings, like the one in Fig. 78). In order to express things in precise terms, the *resolving power* was defined. This is the reciprocal of the minimum angle $\Gamma$ subtended at the eye by two sources at the limit of resolution. However, in practice not $1/\Gamma$ but $\Gamma$ is considered. It would be proper to call $\Gamma$ the *smallest angle resolved* under those conditions. Unfortunately common usage talks about resolving power and employs $\Gamma$, a habit that sometimes gives rise to complicated formulations since one increases when the other decreases.

As our first conclusion we may write

$$\Gamma = k \frac{\lambda}{D}$$

where $k$ is a constant of proportionality. This formula is derived from the theory of diffraction, the radius of a centric's disk being proportional to the ratio $\lambda/D$. For this equation to have real significance, however, the irradiance must be felt in a constant manner by the retina, and to define such constancy when $\lambda$ changes is a rather delicate operation.

On the other hand, the resolving power $1/\Gamma$ depends on the intensity of the sources. This dependence should be represented by a curve that goes down toward zero for both weak and strong intensities. It should go down for high intensities since these lead to a widening of the centrics' disks, partly because of the greater diffusion of the radiation in the retina and humors of the eye, and partly because of the increase in Fechner's fraction. The curve should go down for very low intensities too, since these boost Fechner's fraction considerably, and also since the resolving power must certainly approach zero when the maxima approach the threshold. In fact, when at least one of the maxima drops below the retina's threshold of sensitivity, the resolving power must obviously be zero.

Finally, the sensitivity of the retina clearly exerts an influence, because it supplies the value of the threshold, and at the same time helps to determine the diameter of the disk of the centrics on the retina.

**220.** At this juncture the following comment is in order. For the mind to be informed that there is a minimum between the two maxima of the centrics, there must be a difference between what is carried by the nerve impulses coming from the retinal elements affected by the maxima and minimum, respectively. But individual impulses are all equal. Moreover, the modulation of the frequency may be equal for both these trains of impulses, if the waves involved are pure. Hence, for the two trains to be perceptibly different, the difference must be in the carrier frequency. If the trains of impulses are long, all is well. But if they are short (that is, if the irradiance lasts a brief time, say, a fraction of a second) the difference in frequency may not be noticed, because in so small an interval there is not a difference of even one impulse altogether.

Therefore, the shorter the duration of the irradiance becomes, the greater the difference in irradiance between the maximum and the minimum must be, in order to be felt. In other words, the resolving power $1/\Gamma$ is influenced also by the duration of the irradiance. Excessive shortening of this duration produces effects comparable to those resulting from a drop in the intensity of the sources, that is, a weakening of the resolving power.

Of course, all the agencies that influence these factors likewise affect the resolving power indirectly. For instance, besides the two sources to be resolved, there may be additional and very intense sources of waves in front of the eye. Then the waves that enter the eye are perceptibly diffused by the humors, and thus project an irradiance upon the retina, including the region of the two centrics in question. This diffusion causes a decrease in Fechner's fraction there, and consequently a drop in the resolving power $1/\Gamma$.

**221.** In conclusion, the resolution of two point sources observed through an artificial pupil so narrow that the irradiance projected on the retina by each source occurs in a regular centric is a complex phenomenon, in which energy plays the predominant role. The participants are the source, with the energy of the waves that it emits and with their $\lambda$; the artificial pupil, with its diameter; and the retina, with its sensitive properties, namely, the threshold of sensitivity to radiation, and Fechner's fraction, which in turn is the threshold of sensitivity to differences of irradiation.

In other words, to resolve two sources means to *feel* the difference in irradiance between the maxima and the intervening minimum.

This concept enables us to account for the phenomenon, even if we discard the artificial pupil and expose the eye directly to the action of the waves emitted by the two sources. Instead of the irradiance occurring in a regular centric on the retina, the energy will be distributed in a central maximum and offshoots forming a star around it. The problem becomes more complicated because, the centric's symmetry of revolution being no longer present, the effects are no longer equal in all directions perpendicular to the axis of the eye. Moreover, when the two stars are very close to the limit of resolution, the separation may not be felt in the thickest parts of the two figures on the retina, while being felt at the tips of the thinnest rays perpendicular to the line joining the two maxima. It is difficult to say what the limiting condition ought to be. Yet there is no doubt that in the absence of an artificial pupil, resolution is a function of the same factors as those on which it depended when the pupil was present. For the reasons why the resolution must weaken when the intensity of the sources increases or decreases too much remain the same. Under these conditions too, therefore, resolution is essentially an energy phenomenon, which is at its best when the intensity of the source is suitably related to the sensitive properties of the retina.

**222.** However, an observer is seldom called upon to resolve a pair of point sources, and therefore this case is of greater theoretical than practical interest. Much more interesting, on the other hand, is the resolution of details in an extended source.

In practice the mind "looks at" surrounding objects in order to know them and recognize them for the needs of life. Hence the usefulness of vision consists of perceiving the forms, colors, and locations of objects. The more refined this operation is, the better and more useful vision is.

To see a detail of an object means that the mind has represented that detail in the effigy that it has created. To represent the detail, it has drawn upon the reports reaching it by way of the optic nerve. For these reports to serve the purpose of representing that detail, they must differ from the reports arriving from adjacent points. For these reports to differ from one another, they must be carried by trains of impulses differing in carrier frequency, modulated frequency, or index of modulation. For these trains of impulses to differ, the effects of their respective irradiances on the retina must be *perceptibly different*. To be perceptibly different means that the difference between them must be such that the retina is capable of feeling it. Such a characteristic, consequently, depends either on the energy flux in question or on the sensitive properties of the retina.

**223.** These concepts can be set forth a little more definitely by devising a procedure for describing the action of the radiation on the retina. The elements of the procedure that we shall use are simple and reasonable. Moreover, if they had to undergo some modification, the general course of our reasoning would not change substantially.

Let us imagine, then, that the action of the irradiance on the retina is zero so long as the irradiance remains below a certain value. For in § 201 we took for granted that a threshold of sensitivity exists. On the other hand, a saturation level also must exist. When the energy received is so great that it produces its

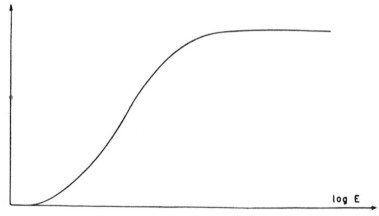

Fig. 83. Effect of radiation on the retina, as a function of the irradiance E

effects on the entire photosensitive substance, clearly any additional energy arriving can produce no further effect. In Fig. 83 the abscissa shows the logarithm of the irradiance $E$, whose effect is indicated by the ordinate. It does not matter in what units the effect is indicated, the purpose of this graph being merely illustrative.

Now take a series of areas $\sigma$, which emit visual waves. Let $\epsilon_1$ be the emittance of one group of these areas, and $\epsilon_2$ the emittance of a second group interdigitated with the first group, as in Fig. 84. Let $\epsilon_1 < \epsilon_2$. The ratio $\epsilon_1/\epsilon_2$ may be regarded as a measure of $c$, the contrast between the two groups of areas.

If an eye receives the waves emitted by these areas, it makes the waves convergent with their centers on the retina. This will, therefore, have areas corresponding to the emitting areas. Let the surface of a retinal area be $\sigma'$. The irradiance $E_1$ of the areas of

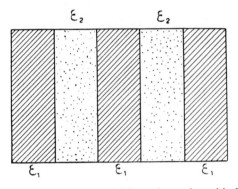

FIG. 84. Field of gray stripes, brighter alternating with darker

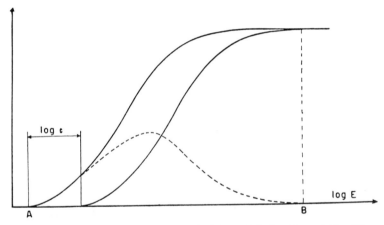

FIG. 85. Effect on the retina of the bright and dark stripes of Fig. 84

emittance $\epsilon_1$ must be related to the irradiance $E_2$ corresponding to the emittance $\epsilon_2$ as follows:

$$\frac{E_1}{E_2} = \frac{\epsilon_1}{\epsilon_2} = c$$

If we now indicate on the axes of Fig. 83 the effects of both groups of areas, we have two identical curves that are separated in the direction parallel to the $x$-axis by a distance equal to $\log E_1 - \log E_2 = \log c$, the logarithm of the contrast (Fig. 85).

If the irradiances on the retinal areas $\sigma'$ are contained within the region between the origin of coordinates and the point $A$, obviously there can be no resolution. For the effects are equal on all the retinal areas involved, and therefore no differences can be felt. The same conclusion must be drawn if the irradiances are greater than those corresponding to the point $B$. Consequently only in the interval of irradiances included between $A$ and $B$ is it possible to have resolution. In exact terms, the perceptible differences are proportional to the differences between the ordinates of the two curves.

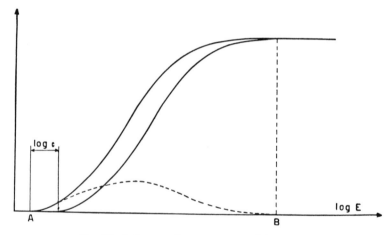

FIG. 86. Influence of contrast on resolving power

Hence the broken curve shows us the movement of the resolving power as a function of the emittance and of the contrast. The presence of the maximum in the middle should be noted, as well as the drop for strong and weak irradiances.

Fig. 86 illustrates how the resolving power is affected by contrast. If $\epsilon_1$ and $\epsilon_2$, the two emittances, are closer together than in the previous case, the logarithm of the contrast $c$ is smaller. The two curves are less distant from each other, and therefore the difference between the effects is smaller.

**224.** The difference between the retinal effects of the two irradiances is shown in Figs. 85 and 86. Resolution occurs when this difference attains a certain minimum. But if the difference is so small that the impulses reaching the brain from the more intense areas do not differ by even one unit from the impulses coming from the less intense areas, then obviously the mind is not informed about the difference. Where the difference between the ordinates is greater, the areas $\sigma'$ may be small and yet the difference between the impulses will still be noticed. On the other hand, where the difference between the ordinates is less, a bigger area $\sigma'$ is required for the difference between the impulses to be felt.

This amounts to saying that if the conditions of emission are good, the details noticed and therefore represented in the effigy are finer. If the emission of the visual waves is either too weak or too strong, the details must be larger, to be noticed and represented in the effigy.

Such excellence or unsuitability of emission can be defined, not in absolute terms, but only in relation to the sensitive properties of the retina concerned. For example, take an eye that has remained for a long while away from light and is, therefore, adapted to darkness. If it is suddenly exposed to intense radiation like that of a sunlit place, it distinguishes nothing because it is operating in circumstances of excessive irradiance. After several minutes, when the retina has adapted itself to the new energy conditions, the resolution becomes very good.

In § 220 we learned about the effects produced by curtailing the duration of the irradiance. Excessive limitation of this period is found to decrease the resolving power of the eye in direct observation.

All the facts mentioned here, as is well known, are in complete agreement with the findings of numerous investigators of the eye's behavior. In particular, the countless measurements of visual acuity by ophthalmologists and students of optics fall into place in the picture drawn above.

Everybody has noticed when reading a newspaper at dusk that as the light grows dim, it becomes more and more difficult to make out the finer print while the larger letters are still easily read, until by moonlight only the name of the newspaper is legible. On the other hand, it is readily recognized that vision is difficult also under conditions of excessive luminosity, usually called *dazzling glare*.

**225.** Let us now go on to examine the acuity of vision through optical systems. The most interesting cases are undoubtedly the telescope and microscope.

From our point of view the action of a telescope is threefold: it increases the angular magnitude of the details of an object, it diminishes the corresponding irradiance on the retina and it introduces an artificial pupil.

With regard to the first action, it was pointed out in § 183 that the angular magnitude of a figure seen by means of a telescope is as many times greater than the corresponding magnitude seen directly (from the center of the objective) as the objective's focal

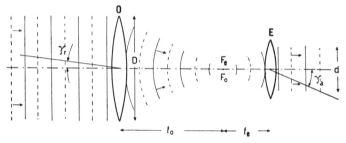

FIG. 87. Geometrical elements of a telescope

length $f_o$ exceeds the focal length $f_e$ of the eyepiece. On the other hand, the diameter $d$ of the waves emerging from the eyepiece (Fig. 87) is smaller than the diameter $D$ of the objective, the ratio being

$$\frac{D}{d} = \frac{f_o}{f_e}$$

$D$ is also called the diameter of the telescope's *entrance pupil*, and $d$ the diameter of the *exit pupil*.

If the angular magnitude of an object seen from the objective (for greater precision we should consider the trigonometric tangent of the angle) is indicated by $\gamma_r$, and $\gamma_a$ stands for the angular magnitude of the corresponding figure seen in the telescope, then

$$\frac{f_o}{f_e} = \frac{D}{d} = \frac{\gamma_a}{\gamma_r}$$

*Real angular magnitude* and *apparent angular magnitude* are the terms for $\gamma_r$ and $\gamma_a$, respectively.

The telescope has also a factor of transmission $\tau$, which is less than unity. Luminous flux is lost because it is reflected from the surfaces of the lenses, absorbed by the glass of the lenses, and diffused by possible obstacles, such as dust, wires, etc.

**226.** We shall now draw up a balance sheet of all these actions.

When the waves of an *ideal* point source reach the eye of an observer through a telescope, the retina is affected by an irradiance whose energy distribution (on the assumption that the instrument is *ideally* perfect) depends on the diameter $d$ of the exit pupil. The instances in which $d$ is very large are not numerous and may be called exceptional. What is found in these instances is obvious. The irradiance on the back of the eye shows a central maximum and a series of offshoots in the form of a star. The effigy created as a result is an irradiation star, as happens with the naked eye. The difference between observation with the naked eye and with a telescope is a matter of energy. In the latter case, the affected area of the retina is struck by radiation that enters through the objective and leaves through the eyepiece, provided of course that the diameter of the exit pupil is no larger than the diameter of the pupil of the eye. In direct vision, on the other hand, only the radiation that enters the pupil of the eye arrives at the retina.

For the sake of simplicity, suppose the exit pupil is equal to the pupil of the eye. Then the intensities of the two radiations are related in the ratio $\tau D^2/d^2$, which ordinarily is much greater than 1. The telescope's star effigy is, therefore, much more brilliant than the naked eye's. With the naked eye, the irradiance may not reach the threshold, while crossing it with the telescope and thereby making *point* sources visible that otherwise would not be visible.

**227.** But if the diameter of the exit pupil is smaller than the diameter of the pupil of the eye, and is actually less than 1 mm, then an artificial pupil is before the eye. A centric is formed on the retina, and the effigy created has the typical structure shown in Fig. 78. The effect of the telescope is, therefore, as follows. The entire flux of energy that passes through the objective and emerges from the eyepiece is distributed in the centric. This is consequently made $\tau D^2/d^2$ times more intense than it would be if a diaphragm of diameter $d$ were put directly in front of the eye. But now the centric's dimensions, in accordance with the rules laid down earlier in this Chapter, determine the angular dimensions of the point source. These angular dimensions are associated with the train of waves emerging from the eyepiece, that is, with the *apparent* field. Hence the dimensions of the optical point source in the *real* field are determined by the *same criteria* as those used for vision through an artificial pupil, the result being divided by the ratio $D/d$.

Yet we must bear in mind that the dimensions of a point source for a given observer are not obtained by supplying him with an artificial pupil of diameter $d$ and then simply dividing the angle thus found by $D/d$. We must remember that the irradiance increases in the ratio $\tau D^2/d^2$. Accordingly, the angular dimensions of an optical point source for a certain eye using an (optically perfect) telescope with entrance pupil of diameter $D$ and exit pupil of diameter $d$ may be said to be $d/D$ times those for the same eye equipped with an artificial pupil of diameter $d$, but operating with a source whose intensity is $\tau D^2/d^2$ times greater than that employed for the telescope.

This conclusion in this form applies also to exit pupils so big that the irradiance on the retina loses the configuration of a centric and assumes that of a star.

It is easy now to go on to the resolving power. An analysis entirely similar to the preceding leads at once to the conclusion that the smallest angle resolved by a certain eye looking through a telescope is $d/D$ times the angle which would be resolved by the same eye with an artificial pupil of diameter $d$ and with a source of waves $\tau D^2/d^2$ times more intense.

**228.** We come now to linear objects. Consider once more an exit pupil so small that the distribution of the energy transmitted to the retina by a point source of waves takes the form of a centric. If the source is linear and subtends an angle $\gamma_r$ at the objective, at the eye behind the eyepiece it subtends an angle $\gamma_a$, equal to $\gamma_r D/d$.

The flux projected by the source upon the objective is proportional to the objective's area, or to $D^2$. Therefore, the flux which enters the eye is proportional to $\tau D^2$. This flux is distributed over a retinal area, which is rectangular in form, one side being equal to the diameter $2r$ of the disk of the centric corresponding to the point source, the other side being proportional to $\gamma_a$. Hence the area is proportional to $\gamma_a$. Consequently the irradiance is proportional to $\tau D^2 / r\gamma_a$.

If this linear source were observed directly through an artificial pupil of diameter $d$, equal to the diameter of the telescope's exit pupil, the eye would receive a flux proportional to $d^2$, which it would distribute over a retinal area proportional to $r'\gamma_r$, $2r'$ being the diameter of the centric's disk for a point source seen through the artificial pupil alone. Accordingly the irradiance is now proportional to $d^2 / r'\gamma_r$, and the ratio of the two irradiances is given by

$$\frac{\tau D^2}{r\gamma_a} \frac{r'\gamma_r}{d^2} = \tau \frac{D}{d} \frac{r'}{r}$$

Separating out the variation in the width of the retinal irradiance as represented by the ratio $r'/r$, we see that now, in the case of the slender source, the irradiance increases, not in the ratio $\tau D^2/d^2$ as previously in the case of the point source, but only in the ratio $\tau D/d$.

**229.** We turn now to an extended source. Instead of considering its length, we must deal with its surface $\sigma$. If $\epsilon$ is its emittance, its intensity is $\epsilon\sigma$. The flux that it transmits to the objective is again proportional to $D^2$, and that which reaches the retina to $\tau D^2$. But the retinal area over which it is spread is proportional to $\gamma_a^2$. Since the effect of diffraction may now be regarded as negligible, the retinal irradiance is proportional to $\tau D^2/\gamma_a^2$.

If the eye looks at this same source through an artificial pupil of diameter $d$, the retinal irradiance is proportional to $d^2/\gamma_r^2$. The ratio between the two irradiances is given by

$$\tau \frac{D^2}{\gamma_a^2} \frac{\gamma_r^2}{d^2} = \tau$$

The ratio is, therefore, constant and less than unity.

**230.** To account for the energy behavior of a telescope, it is very important to distinguish the case of a point source from that of a linear source, and both these cases from that of an extended source. We may thereby explain the following experimental facts. When we gaze at the starry sky at night, we see many more stars with a telescope than with the naked eye. The stars seen in both these ways appear much more intense with the instrument than without it. But when we look at a landscape, i.e., extended objects, the result is entirely different, for there is no notable advantage.

However, the situation is not so simple as a schematic theory, like the one set forth here, may suggest. The extended object considered in the calculations of § 229 had a uniform structure, since the distribution of the energy on the retina was computed by dividing the total flux by the irradiated area. In practice the telescope is used to see the details of objects better. Hence objects with fine details are observed, and as such come under the heading of linear sources, if not actually point sources.

The total flux entering a telescope, when pointed at an extended source too, is proportional to $D^2$, whereas when we look with the naked eye, the flux is proportional to $d^2$. This fact also is not without effect. To be sure, in the former case the flux is distributed over a greater retinal area, so that the average flux is practically unchanged (or if it varies at all, it decreases a little as a result of $\tau$). But we know that the retina does not have the same sensitivity throughout, the parafoveal region being the most sensitive. Moreover, since the mind receives a larger number of reports, it feels that more radiation is arriving from the outside when we look through the telescope. Hence at night anybody looking through a telescope even at extended objects creates effigies brighter than those fashioned with the naked eye, despite $\tau$.

**231.** In conclusion, we may say that the telescope has the following three effects. The first is geometric, so to speak, because it multiplies by a certain factor the angular magnitude subtended by objects seen in direct vision. The second effect is photometric, since it modifies the irradiance on the back of the eye. The last effect may be called optical, inasmuch as it enables the eye to resolve smaller details of an object than it would have resolved in direct vision. In each of these functions the telescope may be regarded as a device auxiliary to the eye, that is, as a device capable of heightening the powers of the eye.

**232.** A similar analysis may be made for the microscope. The difference in structure between the two instruments corresponds to the difference in the distance from the object to the objective. But the function of both devices is to enable the eye of the observer to resolve finer details of the object than the eye would be able to resolve in looking directly at the object.

In direct vision the object would subtend a highly variable angle depending on the distance from the object to the eye, the distance involved being of the order of magnitude of some inches. It is, therefore, difficult to compare the angular magnitude of the object viewed directly with the angular magnitude of the figure seen with the aid of a microscope. The task has been facilitated by the conventional practice of putting the object 25 cm from the eye in direct vision.

Turning back to Fig. 76, we see at once that $\dfrac{y'}{f_e}$ is the angle subtended by the effigy at the eye, while $\dfrac{y}{25}$ is the angle subtended by the object at the eye in direct vision. Hence the increase that the microscope produces in the angular magnitude is given by

$$\frac{y'}{f_e} : \frac{y}{25} = \frac{y'}{y}\,\frac{25}{f_e}$$

But the ratio $y'/y$ is equal to $x'/x$ or, with the simplifications adopted in § 188, to $\dfrac{\Delta}{f_o}$. On the basis of the aforementioned convention, the increase in the angular magnitude is consequently given by

$$\frac{25\Delta}{f_e f_o}$$

**233.** Is this angular magnification associated with a proportional diminution of the dimensions of a point source and of the smallest resolvable angle? In order to answer this question we must examine the movement of energy in the whole arrangement; that is, we must see what happens to the exit pupil. If photometric conditions in the eye were identical and invariable, comparison of the smallest angle resolvable with and without the microscope would be simple. If the smallest angle resolved in direct vision is indicated by $\Gamma$, the smallest angle $\Gamma_m$ resolved by observation with the microscope would be

$$\Gamma_m = \frac{\Gamma}{\dfrac{25\Delta}{f_o f_e}} = \Gamma \frac{f_o f_e}{25\Delta}$$

Call the pupil in direct vision $d_o$, and the exit pupil of the microscope $d$. On the assumption that the energy distributed on the back of the eye takes the form of a regular centric in both cases, the above value of $\Gamma_m$ is multiplied by the ratio $d_o/d$. For the dimensions of the centrics on the retina in the two cases are related to each other inversely as the diameters of the pupils. Designating the smallest resolvable angle thus corrected as $\Gamma'_m$, we have

$$\Gamma'_m = \Gamma_m \frac{d_o}{d} = \Gamma \frac{f_o f_e}{25\Delta} \frac{d_o}{d}$$

But obviously

$$\frac{d}{D} = \frac{f_e}{\Delta}$$

Therefore

$$\Gamma'_m = \Gamma \frac{f_o}{D} \frac{d_o}{25} = \Gamma \frac{1}{2 \sin \alpha} \frac{d_o}{25}$$

where $\alpha$ is half the angular aperture of the microscope's objective as seen from a point on the object.

**234.** The entire advantage, therefore, lies in making $\alpha$ as big as possible, although of course $\sin \alpha$ can never exceed 1. Hence it would evidently follow that when the angle $\alpha$, which is the only factor at our disposal, reaches its maximum extent, everything humanly possible has been obtained from the microscope.

But in truth the situation is much more complicated. The value of $\Gamma'_m$ cannot be related to $\Gamma_m$ modified simply by the ratio of $d_o$ to $d$, the diameters of the pupils. To be sure, in general $d$ is really small enough to distribute the energy on the retina in the form of regular centrics. But in the case of the unobstructed pupil $d_o$ there certainly are no centrics; and since stars reappear, nothing can be said with assurance and the matter assumes a personal character. This lack of generality is of no great consequence, however, because in this discussion the values $d_o$ and 25 cm are, more than anything else, conventional.

The topic of greatest consequence, on the other hand, concerns the distribution of the energy. The flux that enters the objective of the microscope later spreads out over a surface proportional to the square of the apparent angle subtended by the effigy. Since this angle is rather large, the irradiance on the retina would drop to values so low that it actually would not even reach the threshold of sensitivity, if the microscope were placed directly over an object under conditions of ordinary illumination.

It has been possible to obtain extensive results from the instrument, in fact results bordering on the maximum attainable, by using a *condenser* to concentrate an extremely vigorous flux of radiation on the object under examination so as to raise its emittance to very high values. It nevertheless remains the responsibility of the skillful observer in each instance to choose the illumination conditions best suited to the magnifying power of the instrument, the sensitivity of his own eye, and the transparency of the object, in order to obtain in every situation the strongest contrast and therefore the most effective resolution.

In general, with instruments of weak magnification (100 to 200 times) the illumination must be cut down below the maximum to avoid reducing the resolution on account of dazzling glare. On the other hand, when the magnification rises to the upper ranges (1000 to 2000 times) luminous sources of the greatest emittance and the mightiest condensers are needed to obtain the finest results in resolution, and sometimes it is felt that more could be achieved if it were possible to illuminate the object even more intensely.

**235.** Another instrument that deserves to be considered at this time is the spectroscope.

As was remarked in § 165, when waves emitted by a linear source pass through a refracting prism, whose refracting edge is parallel to the source, and are received by an eye, it sees a line spectrum, namely, many linear figures like the source, of different hue, arranged parallel to one another, each of them corresponding to a different wave length of the waves emitted by the source.

Now suppose the source emits two waves, of wave length $\lambda$ and

$\lambda + d\lambda$, with $d\lambda$ very small. This smallness allows the hues of the two corresponding lines seen by the observer to be sensibly equal, and therefore color no longer constitutes a distinguishing characteristic. The difference between the two wave lengths is known to the observer only from the fact that there are two lines, in other words, that the radiation on the retina is distributed along two linear parallel zones.

Of course the bigger $d\lambda$ is, the farther apart the two lines are, because the prism disperses the two waves more widely. On the other hand, if $d\lambda$ tends toward zero, the distance between the two lines likewise tends toward zero. Hence there is a minimum value of $d\lambda$ which can be perceived in this manner. For when the two lines are so little distant from each other that they seem to the observer to be only one, the existence of $d\lambda$ is no longer made known. The problem then is to determine this minimum value of $d\lambda$ and the factors on which it depends.

**236.** The way to solve this problem is clear. An angle $\delta$ of deviation by the prism corresponds to every $\lambda$ emitted by the source. When this emits two waves, on emerging from the prism they form an angle $d\delta$ between them. This angle must be apprehended by a suitable means of observation. The more powerful the means, the smaller will the least perceptible $d\delta$ be, and the smaller the $d\lambda$ that can be detected by this means.

Since the angle $\delta$ depends on the angle of incidence, it is necessary at the outset to make sure that all wave elements of the same $\lambda$ reach the prism's face of entrance at the same angle of incidence. Since this face is plane, the incident wave also must be plane. To accomplish this result, the source is put either at a great distance from the prism (a very difficult thing to do outside of astronomy) or in the focal plane of a converging optical system. The latter is the arrangement generally adopted (Fig. 88). The radiation that is to be analyzed illuminates a slit $S$ at the focus of an objective $O$, the combined slit and objective being called a *collimator*, $C$.

The waves thus enter the prism at a well-defined angle of incidence. On emerging, those of wave length $\lambda$ form with those of wave length $\lambda + d\lambda$ an angle $d\delta$ (§ 318) expressed by

$$d\delta = \frac{b}{D} dn$$

where *dn* is the corresponding variation in the prism's refractive index, *b* is the useful length of its base, and *D* is the breadth of the

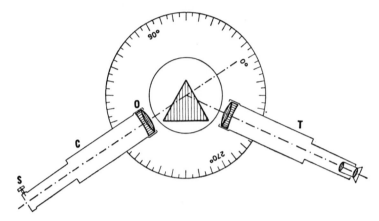

Fig. 88. Diagram of a spectroscope

wave front emerging from the prism, this breadth being measured in the direction perpendicular to the edge of the prism.

**237.** These waves could be sent directly into the eye of the observer. Then the minimum $d\delta$ would be equal to the smallest resolvable angle $\Gamma$ characteristic of that eye for a linear source of that given intensity and $\lambda$. In that case we could write

$$d\delta = \frac{b}{D} \frac{dn}{d\lambda} d\lambda = \Gamma$$

Hence we could deduce $d\lambda$ by recognizing that $\frac{dn}{d\lambda}$ is the so-called dispersion by the glass of the prism for the interval $d\lambda$ of the wave lengths under consideration, and is therefore a specific characteristic of the glass. But clearly this is not the best procedure, since we have at our disposal a means of resolving small angles, namely, the telescope. It is the telescope (*T* in Fig. 88) which is used to convey the waves to the eye of the observer.

Obviously, however, not any sort of telescope can be employed, because the diameter of its objective is governed by *D*, the breadth of the wave emerging from the prism. If the diameter of the objective were bigger, the region beyond *D* would remain unused, as if it were not there; and if the diameter were smaller, a piece of the prism would remain unused.

Assuming then that the diameter of the objective is exactly equal to $D$, we must choose the focal lengths of the objective and eyepiece so as to find the most satisfactory exit pupil in accordance with the familiar equation

$$\frac{D}{d} = \frac{f_o}{f_e}$$

The terms in which this question is handled are clear. We have

$$\frac{D}{d} = \frac{\gamma_a}{\gamma_r}$$

and in this case $\gamma_r$ is precisely $d\delta$. If the smallest angle resolvable by the eye remained constant, the entire advantage would be obtained by making the ratio $D/d$ as big as possible. But we know that as this ratio rises, $d$ steadily declines, and hence the smallest resolvable angle increases. This increase occurs in that complicated manner which we examined in § 227, where we saw that it is a function also of the intensity of the source.

Consequently much depends on this intensity. If it is very great, the aforementioned ratio can be pushed up and finer results obtained. If the intensity is very low, the ratio must be limited so as to put the eye in better observational conditions. There are no fixed conditions which guarantee a given resolution.

**238.** We have been discussing a prismatic spectroscope. But our analysis applies to any other type of spectroscope equipped with dispersion apparatus, like a line grating, echelon grating, or Lummer plate. To confirm what was said about the spectroscope I may cite the long and conclusive experience of comparisons between the prism and the line grating. The latter's angular dispersion is so high by contrast with the former's that apparently the grating ought to replace the prism in all applications of the spectroscope. Unfortunately the grating, with its division of the incident wave into numerous diffracted waves, has so low an energy yield as to be useless in all the cases (and they are not few) in which the radiations emitted by sources are not very intense. Today, after spectroscopes have been employed for about a century, the principal place in practice is still held by the prismatic type.

**239.** Now that we have come this far, I believe it is time to take notice of the disappointment undoubtedly felt by the reader who finds himself faced by a series of conclusions that are vague, extraordinarily intricate, and above all hardly inescapable. If he has studied what the standard works say about the instruments just reviewed, he there encountered very simple, precise, and unusually stimulating conclusions. He will wonder whether it is worthwhile replacing these laws with the nebulous and complex considerations so laboriously put together in the preceding pages. This is not the first occasion on which I have been obliged to answer this kind of question.

Some decades ago almost all students of optics spoke the language of geometry and found that everything went marvelously well. That had been the situation for centuries. There were some strange individuals who derived deep satisfaction from mixing matters up by talking a new language, the language of waves. The numerous devotees of geometrical optics asked themselves, Is it worth the trouble? Geometrical optics is so well organized, so unified, so complete, and at the same time so easy, accessible, and clear. What advantage is there in superseding it by so much more involved and difficult a scheme that complicates the calculations with new parameters?

The advantage was there, and it asserted itself with steadily increasing force until nowadays the wave language has become as common as the geometrical, and perhaps more so. The advantage consisted of the fact that the wave mechanism accounted for a vast group of phenomena, such as interference, diffraction, and polarization, which geometrical optics could not incorporate in its picture, and which its devotees therefore found it very convenient either not to know or to ignore.

But in practice those who experimented in laboratories and worked on technical applications found it very useful to develop the study of the phenomena that the devotees of geometrical optics did not want to consider because these phenomena were not readily reconciled with their idol, the luminous ray. And so wave optics became easy, clear, accessible, and yielded exact, simple, thought-provoking laws.

**240.** Maybe these laws were too thought-provoking. They have followers today who should rather be called unconditional admirers, who accept these laws while closing their eyes to experience. The earlier episode is being repeated in another form. Once more it is essential to examine these beautiful, simple, and exact laws with a critical mind and in the light of experience. Then we shall decide whether or not it is advantageous to change our course.

As regards direct vision, it has been known for a long time that there is a limit to the fineness of the details that an eye can distinguish. The reason for this limitation has been ascribed, on the one hand, by physiologists to the cellular structure of the retina and, on the other hand, by physicists to the wave structure of the radiation.

The reasoning of the former group appears very clear and logical. In the fovea there are many cones, one right next to another. If there are two point sources of waves so close to each other that the centers of their respective waves in the eye fall on the same cone, the sources are seen as a single point. If the two centers fall on two contiguous cones, the sources are seen as a length. If the centers fall on two cones separated by at least one nonstimulated cone, the sources are seen as distinct. Therefore, the limiting angle of resolution is a consequence of the cellular structure of the retina, and is a constant of the eye. Its value is calculated by taking the distance between the centers of two cones separated by a third cone, and dividing that distance of 6 $\mu$ by the distance of 20 mm from the second nodal point of the eye to the retina. The quotient comes out $3 \times 10^{-4}$ rad $= 1'$. Measurements of visual acuity often give this value. The physiological explanation has been adjudged to be in conformity with experience, and the angle of $1'$ has been regarded as a basic element in the problems of vision.

**241.** Although this style of reasoning is still accepted and approved by many students today, it is open to very serious objections.

First of all, it never happens that the radiation emitted by a point source is concentrated so as to affect only a single cone. Particularly when the pupil is unobstructed (that is, when no artificial pupils are put in front of it) there are irradiances in the form of a star, which are more conspicuous the more intense the source is. To have very compressed concentrations of irradiance on the retina, we must go down to extremely low intensities; but then the pupil dilates and the aberrations of the eye are more marked, so that the affected retinal zone is bigger, for these reasons. Consequently the basic assumption on which the physiologists' argument rests is faulty.

Furthermore, the conclusion that the smallest angle resolvable by the naked eye is a constant is itself contrary to experience. This angle has been shown by all experimenters to be a function of the intensity of the source. The angle may be represented in a graph as a function of the logarithm of the intensity, with a central maximum and continuously descending slopes for both increasing and decreasing intensities.

Again, when we turn to observations made not of point sources but of details of extended objects, contrast between the light and dark parts makes its influence felt, as does also the duration of the observation. How these factors can be explained by the supporters of the physiological theory is not clear.

Besides, in the calculation leading to the famous 1' the dimensions of the cones do not correspond to the dimensions of the cones found in the center of the fovea, the latter dimensions being less than half the former.

Moreover, not all individuals have 1' as their smallest resolvable angle. Some have a smaller angle, some a larger, even when the optical system of the eye is brought to maximum efficiency by means of corrective eyeglasses. But even more difficult to reconcile with the physiological theory is the fact that in the same eye the angle itself changes continuously with fatigue, for example after a prolonged observation, and with many other factors, as for instance the percentage of oxygen in the air breathed by the observer.

It is not at all evident how this whole mass of phenomena can be brought into agreement with the idea that the smallest angle resolvable by the naked eye is a constant, linked directly to the cellular structure of the retina.

**242.** Equally awkward is the situation of the other idea which, however, has lost much ground nowadays, that the limit of resolution by the naked eye is a function of the diffraction phenomena due to the wave structure of the radiation. In other words, the waves that enter the eye are limited by a diaphragm 2 mm in diameter, and therefore should show at their center on the retina a distribution in the form of a centric. Its disk has an angular diameter (§ 328) expressed by

$$2\gamma = 2\frac{1.22\lambda}{D} = 1.22 \times 5 \times 10^{-4} = 6 \times 10^{-4} \quad \text{rad}$$

$D$ being equal to 2 mm. The diameter of the disk, then, is 2′. Hence, in order to find $3 \times 10^{-4}$ rad, or 1′, as the smallest resolvable angle, we need only think that two centrics are seen resolved when their centers are separated by a distance equal to the radius of the central disk.

All would be well were it not for the very serious objection that, with the pupil 2 mm in diameter, in almost all eyes the waves are concentrated, not in a centric, but in a star. Moreover, unless intervention by the retina in the phenomenon is introduced, it is not clear how it is possible to explain most of the experimental facts mentioned in § 241, such as the dependence of the smallest resolvable angle on the duration of the observation, on the fatigue of the observer, and on the air that he breathes. But there are other specific considerations that raise difficulties for this theory. The smallest resolvable angle should decrease when the pupil's diameter $D$ increases, and should also vary proportionately with $\lambda$. Neither of these things happens. Disturbing interferences by imperfections and aberrations may be invoked in this connection. But in fact no such interference is confirmed, except when a very small artificial pupil produces on the retina an irradiance in the form of a centric.

The only effective part of this theory is the argument that it provides against the physiological theory. What is asserted with certainty by the radiation theory is that the distribution of energy on the back of the eye is such that it can never affect only a single cone, even if the waves come from an extraordinarily fine point source.

**243.** The idea that the smallest angle resolvable in direct vision is a constant and is exactly 1′, and that it is in marvelous agreement with the structures of the retina and the radiation, is indeed beautiful, stimulating, and enticing. We may readily understand why this idea was received with such warmth and confidence by all interested students, including physiologists.

But unfortunately it cannot be said to correspond to reality. The smallest angle resolvable in direct vision is not a constant. It is a function of the form of the source, of the structure of the radiation, but above all, of the energy emitted by the source and received by the eye, and of the sensitive properties of the retina.

**244.** Now let us go on to the telescope. At first its function was conceived to be that of a special eyeglass for distant vision. The angles $\gamma_a$ of the apparent field having been found to be larger than the corresponding angles $\gamma_r$ of the real field, the ratio $I = \gamma_a/\gamma_r$ was called the magnification, and the telescope was said to *magnify* according to the ratio $I$. Then, since the smallest angle resolvable by the eye in the apparent field was 1′, that in the real field had to be $1'/I$. Hence it was necessary to be able to reduce the smallest resolvable angle at will, provided the magnification was adequate. Since $\gamma_a/\gamma_r = f_o/f_e$, increasing the focal length of the objective while shortening that of the eyepiece was enough to achieve resolution of ever smaller angles.

Of course experience responded to these projects with a plain "No." In those days the excuse for the failure was found in the aberrations and imperfections of the lenses. But this excuse vanished when excellent reflecting telescopes with parabolic mirrors eliminated all the aberrations and brought the imperfections below optical tolerances. The astronomer William Herschel at the beginning of the last century arrived at this result and naturally observed the centric. Then the conclusion had to be drawn that the problem was complicated by diffraction, and that the smallest resolvable angle was limited by the dimensions of the centric. For as the magnification increased, the apparent angular distance between the centers of the centrics increased too, but the angular magnitude of the centrics themselves also increased, and the separation did not in-

crease. The problem thus became to obtain centrics of the smallest
possible diameter while maintaining the magnification.

**245.** But meanwhile the period in which the eye and the observer were completely banished from these matters was in full
swing, and therefore the telescope was considered by itself. The
eyepiece was regarded as an accessory that permitted the eye to see

Fig. 89. Pair of centrics at the limit of resolution, in Lord Rayleigh's
sense of the term

at very close range the real image formed by the objective. Hence
the problem of the telescope in substance was posed as follows. A
real image of external objects is formed by the objective in (or near)
its focal plane. The structure of this image is not a perfectly faithful
reproduction of the structure of the objects, because diffraction of
the radiation hides details smaller than a certain limit. The question, therefore, is to determine exactly what the effect of diffraction
consists of, and on what factors it depends.

Attention was then directed to the double star. Each of the pair
has for its real image (Fig. 79) a centric whose disk is subtended at

the center of the objective by the angle

$$2\gamma = 2\,\frac{1.22\lambda}{D}$$

When the sources approach each other, the centrics also draw closer together and overlap more and more until they merge and seem to be only a single centric.

Lord Rayleigh proposed to take as the limit of separation between two centrics that corresponding to the overlapping by each disk of half of the other (Fig. 89). The smallest angle resolvable with a telescope thereby remained fixed as

$$\Gamma = \frac{1.22\lambda}{D}$$

**246.** This equation has been in existence for seventy-five years, during which time it has been deemed one of the foundations of instrumental optics. Universally known as "Lord Rayleigh's rule," it is indeed thought-provoking by reason of its simplicity and the significance of the parameters which it contains. Only one, $\lambda$, represents the source, and it is the fundamental element of wave optics. Only one, $D$, represents the instrument, and it acts as the *base* of the observation in the surveyor's sense, somewhat like the base of Kepler's telemetric triangle. The two parameters appear in the first power. The rule could not be simpler.

It gave rise to a highly important and most agreeable concept. The smallest resolvable angle was a purely physical function of the source and of the instrument, independently of the observer. Moreover, the source intervened through a practically constant parameter, especially if the observation was to be performed with the daytime radiation, which includes all the visual $\lambda$'s. Hence the angle $\Gamma$ became an exclusive property of the telescope, and as a consequence the *resolving power of the telescope* was defined. This idea is still generally known and accepted, not only in theory but also in the technical and commercial domain, where it is taken as a yardstick for measuring the power of instruments.

Among the other properties of this rule, of particular interest is its independence of the telescope's magnification and of the source's intensity.

**247.** It is undoubtedly a wonderful rule. But even more wonderful is how it has been accepted by the entire optical world and used with blind faith for three quarters of a century, when experience constantly revealed its falsity, above all its conceptual falsity.

Anyone looking back at the way in which the rule was enunciated would be amazed by the little care employed. Yet this unconcern is understandable when it is realized that Lord Rayleigh proposed the rule mainly for its simplicity and convenience, without attaching special importance to it. Hence he did not take the trouble to give it the experimental basis that would have been necessary for a proposition destined from birth to become a fundamental law of optics. Furthermore, in the few experiments he did perform and that he described with great objectivity and precision, when he thought he was demonstrating the validity of the law, the proof was already present that it was not always valid, because it was affected by the energy conditions of the experiment. As Lord Rayleigh himself remarked, "That bright stars give larger disks than faint stars is well known to practical observers." To measure the resolving power of a telescope in radiation of a definite wave length, he used a wire grating or a piece of common gauze as the test object, and illuminated it with a sodium flame. He carried out the measurements by fitting various diaphragms in front of the telescope's objective for the purpose of ascertaining how the smallest angle resolved changed as a function of $D$. He found that the law of inverse proportionality did not hold for the smaller diaphragms. This failure meant that under certain conditions, already present among the very few investigated by the first experimenter, the law did not correspond to experience. But when he assisted the sodium flame with a jet of oxygen, the measurements turned out as prescribed by the rule, for the smaller diaphragms too.

**248.** This style of reasoning is highly significant. When a rule is not verified under certain conditions, instead of concluding that it is not always valid, the conditions are modified to make it come out right.

This method was applied on a large scale also in later years. For the problem often came up of ascertaining the resolving power of a telescope. The reasons were not only purely theoretical but also practical. Industrial products had to be accepted or rejected, and the disposition of large sums of money depended on the outcome. What was the proper procedure for determining the smallest resolvable angle? There was only one way. Use the telescope to look at a test object, whose fineness was varied until its details were no longer distinguishable. It was thus possible to find out what were the smallest dimensions of the details resolved.

However, there was the eye of the observer that could falsify the measurements. There were the conditions of illumination that could be either too dim or too intense. And there were various other factors that could affect the issue, such as fatigue, haste, and worry.

Naturally the utmost care was taken to avoid *disturbances*, to choose *good* conditions of illumination, and to select a keen-sighted inspector, who was unhurried, unworried, and well rested. With all these precautions a number was obtained that was attributed to the telescope, as if the observer counted for nothing. Since the number was found only with all these precautions (for otherwise it would not have been necessary to take them) when it happened that another observer under other conditions of illumination or of time or of anything else did not happen to get the same result, the fault lay, of course, with the new observer.

**249.** It took time to locate the defect in this manner of reasoning. For the origin of the error resided in the initial propositions, namely, in the concept that the so-called *aerial image* in the focal plane of the objective was a physical entity, itself endowed with a certain fineness, that is, with certain details. Had the question been asked by what procedure it was possible to ascertain the dimensions of the finest details in the aerial image, everybody would have found it natural to reply, "Look at them, of course with a very powerful microscope." For everybody was convinced that to look with a microscope meant to discover the construction of the object or figure placed before the microscope's objective.

Instead, to ask how it was possible to determine the finest details in the aerial image meant to pose a very perplexing problem. For to look at the details with an eyepiece or with a microscope (the latter being basically also an eyepiece, with an objective in front of it) simply means to put together a comprehensive optical system (consisting of the objective of the telescope, possibly the objective of the microscope, the eyepiece, and the optical system of the eye), which conveys to the retina the waves emitted by the object. What is seen is that which is derived from the impression of the retinal elements and from the entire subsequent physiologico-psychological process. What is found in the focal plane of the telescope's objective is not seen at all, just as what is found in the plane of the real image formed by the objective of the microscope, when that instrument is introduced, is not seen. Only that is seen which acts on the retina, and it is seen only as it is interpreted by the psychological mechanism.

**250.** But to draw this conclusion means to abolish the *resolving power of the telescope* outright, and together with it Lord Rayleigh's rule in its very essence. It means to destroy in its entirety the simple, precise, and stimulating physical construction, in order ineluctably to reach the new way of viewing the matter. In other words, the resolving power of the telescope does not exist. What exists is the resolving power of the retina. The telescope modifies the distribution of the waves directed toward the eye, so that the latter's resolving capacities may be better utilized. Anybody wishing to ascertain what advantage is conferred by the telescope in its function as thus understood will arrive at precisely the propositions expounded in the new manner on the preceding pages.

Fig. 90 is the best demonstration of how Lord Rayleigh's rule compares with the new theory. The ordinate shows the resolving power, expressed in arbitrary units. The abscissa gives the logarithm of the area of the exit pupil of a telescope with which the same observer looked at a pair of point sources emitting waves of a pure wave length λ. The objective of the telescope was always the same and had a constant diameter. Variation in the exit pupil was secured by changing the eyepiece, thereby altering the magnification.

Under these conditions, according to Lord Rayleigh's rule, the resolving power should have been constant, maintaining the value designated by the straight line parallel to the x-axis and marked "old theoretical value." The curves in the Figure show the experimental results. Each of the curves, numbered from 1 to 7, indicates a series of measurements carried out while the intensity of the sources was kept constant. When it was decreased by putting a steadily thicker photometric wedge in front of the sources, the results

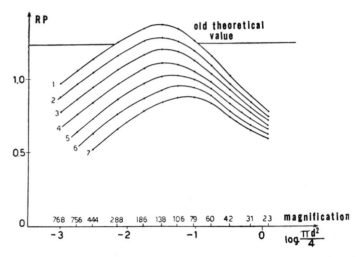

Fig. 90. Resolving power of a telescope as a function of the intensity of the double star under observation and of the logarithm of the area of the exit pupil, while the entrance pupil remained constant

moved from a curve to the next one below it in the increasing order of the adjoining numbers.

**251.** If experience shows us so clearly that resolving power does not follow a simple rule, we should decide whether we want to engage in sheer speculation about what would occur if Lord Rayleigh's rule were verified, or to try to find out what actually happens. If we choose the latter course, we must proceed in the new direction pointed out here.

For Lord Rayleigh's rule to have a meaning corresponding to reality, resolution would have to be a purely objective phenomenon, and also a purely *formal* phenomenon (that is, a function of the form of the media traversed by the waves) and therefore independent of the energy involved. On the contrary, resolution is essentially an *energy* phenomenon, and is strictly subjective, being connected with the sensitive properties of the observer.

These new concepts fundamentally alter the ideas about resolving power. They change its seat, make its mechanism much more complex, but permit us to observe experimental facts with far greater mastery over them.

**252.** What has been said for the telescope may be repeated as a whole for the microscope, with slight variations in form. The *resolving power of the microscope*, like that of the telescope, does not exist. The microscope too is an instrument that modifies the course of the waves emitted by the object, so that the resolving power of the observer's eye may be better utilized. But the microscope has another function, which acquires its proper prominence only in the framework of the new ideas. By means of the source of waves and the condenser, the microscope regulates the photometric circumstances so as to produce the best conditions for resolution by the observer's eye.

The standard equation

$$s = \frac{\lambda}{2n \sin \alpha}$$

which gives the smallest length $s$ of the details at the limit of resolution, has the same value for the microscope as Lord Rayleigh's rule has for the telescope. Moreover, the two equations are the same in substance, and differ in form in order to take account of the different ways in which the two instruments are used.

**253.** The spectroscope, although far less important, is nevertheless interesting in this regard.

Just as the resolving power of the telescope does not exist, so there can be no resolving power of the spectroscope, which in the last analysis functions by means of a telescope.

But it would be well to review the essentials of the standard discussion. Recalling the notation used in § 236, we may equate $d\delta$, the smallest observable difference in deviation, with $\lambda/D$, the value given by Lord Rayleigh's rule, except for the coefficient 1.22. For with this coefficient the value applies to circular pupils, whereas in spectroscopes the entrance pupil of the telescope generally is rectangular. In this context $D$ stands for the breadth of the wave front emerging from the prism and measured in the direction perpendicular to the prism's edge. Then

$$d\delta = \frac{b}{D} dn = \frac{\lambda}{D}$$

whence

$$\lambda = b \, dn$$

and

$$\frac{\lambda}{d\lambda} = b \frac{dn}{d\lambda}$$

The fraction $\lambda/d\lambda$ is taken as the measure of the resolving power of the spectroscope. Once the prism has been set in place, $dn/d\lambda$ is constant. Therefore, the resolving power depends on only one factor, $b$, the base of the prism. This is an equation incomparable in comprehensiveness, simplicity, and elegance. Hence there was talk about the *resolving power of a prism.*

But the properties of resolution have never been enjoyed by the prism, nor by the line grating, echelon grating, nor Lummer plate. These devices are only means of dispersion. They give different deviations to waves of different wave length.

Had it been possible to attribute resolving power to the telescope, as was then done, the correct expression would have been *resolving power of a prismatic spectroscope.* As I showed decades ago, if the telescope is replaced by an interferometer, the spectroscope's resolving power (energy permitting) can even be doubled. That power is, therefore, not a characteristic of the prism.

**254.** We must now recognize that it is not a characteristic of the prismatic spectroscope either, but of the observer equipped with a prismatic spectroscope. Resolving power depends on the intensity of the source and on the structure of the prism, of the telescope, and of the observer's eye. Thus, for example, the two waves to be resolved may be so weak that when they reach the retina, their effect is not enough to make it feel the difference between the maxima and the minimum, despite the whole base of the prism and the objective of the telescope. In that case the observer does not say that he sees two distinct lines, but he sees only one or even none.

There may also be a situation of the following kind. A source emits two very weak waves. A spectroscope with a prism having a base of 10 cm, and with the thick objectives necessarily accompanying such a prism, absorbs so high a percentage of the radiation along the way that what finally reaches the observer's retina does not permit him to feel the difference between the maxima and the minimum, and therefore he does not resolve the two $\lambda$'s.

The same observer, using a spectroscope with a prism whose base is 2 cm and with appropriately thin objectives, receives such intense stimuli on his retina that he notices the double structure and infers that there were two $\lambda$'s. Therefore the observer equipped with a spectroscope having a prism whose base is 2 cm has better resolution than the same observer using a spectroscope with a prism whose base is 10 cm.

The accepted theory asserts the contrary. But instances of this type occur fairly frequently. That is why I was led to formulate a new theory.

# CHAPTER VI

# *The Optical Image*

**255.** Looking back over what has been said, we may distinguish the following three entities:

(*a*) the material *object*, which we regarded as consisting of atoms; as such, for our purpose it acts as a *source of waves;*

(*b*) the *effigy*, which is created by the mind of the observer, either on his own initiative as in a dream, or on the basis of the information reaching his mind by way of the action of the waves on his eyes, or by way of the muscles, or in some other way; effigies are bright and colored figures;

(*c*) the optical *image*, which may be considered the locus of the centers of the waves emerging from an optical system.

As the conclusion of our discussion, we should say that *seeing* means creating an effigy and placing it in a portion of the space in front of us. From the point of view of the visual process there is no distinction between the case in which the waves reach the eyes directly from the object and the case in which the waves undergo a deviation or deformation along the way, as happens when they encounter an optical system. The difference between the two cases may be apprehended by checking with the senses other than sight (generally touch) or as a result of information arriving by some other route. The essential contrast is that in direct vision the object almost always coincides with the effigy, especially if the object is familiar and not very far away from the observer; whereas, when the waves are modified by an optical system along the way, the object is almost never found where the effigy is placed, or at any rate the shape and dimensions of the object differ from those of the effigy.

**256.** When an observer creates an effigy and places it so that it coincides with the object (as may be verified, for instance, by touch) in ordinary language he is said *to see the object*. We too may adopt this phrase, even though it implies ideas that take too much for granted. "To see an object" carries a little too far the implication that what is seen faithfully reproduces what physically exists. The quoted words quietly suggest the conception that the object is as it is seen. This proposition is not true nor can it be true. Vast amounts of research on the mechanism of vision have resulted in the incontrovertible demonstration that the figure seen is created by the mind of the observer. He assigns it a form, brightness, hue, and saturation on the basis of the functioning of his eyes, muscles, and mental faculties, including his memory of the experience and his imagination. He assigns it these traits in so personal a manner that even with the best will in the world he can indicate practically none of them to anybody who asks him about them.

Therefore, to say that the effigy created is *equal* to the object, even when touch and the other senses and means of investigation offer all the confirmation desired, is still a highly debatable statement. For the object, as an aggregate of atoms all consisting of protons, electrons, and other elementary particles, cannot be equal to a figure that has a continuous and stationary structure, and that is bright and colored, whereas the object cannot be bright and colored.

Consequently, the words "to see an object" have meaning only if they are considered to be a conventional expression, more convenient than the one conveying the actual content, namely, *to place the* (best possible) *effigy so that it corresponds to the object* that caused it to be created.

**257.** The phrase "to see an image" is also common. Here we are of course concerned with the optical image, which was defined in § 255 as the locus of the centers of the waves emerging from an optical system.

According to what was said in § 256, "to see an image" should mean to create a corresponding effigy and place it where the image is. But this process, although apparently identical with the other in

its physical and physiological technique, acquires manifestly different characteristics in the psychological phase. For very often the
optical image occurs in a position in which an object could never be
found. Thus reports are contributed that profoundly alter the way
in which the information received along the optical route is translated into the final effigy. Hence of all the possible cases, those in
which an optical image is seen are rare.

It has been my constant aim in the preceding Chapters to make
this conclusion quite clear. The effigy created by an observer employing an optical system is seldom placed in the position of the corresponding image. I have repeatedly emphasized this remark
because seventeenth-century optics started with the assumption
that images are seen. It utilized this assumption, verified in a tiny
number of cases, to attain the goal that it had set before itself,
namely, ousting the eye from the position that belongs to it in vision
by means of optical instruments. On the other hand, I have sought
to show that, for the purpose of a thorough investigation of vision
by means of optical instruments, it is indispensable to restore the eye
to its proper place. For vision, whether it is direct or uses optical
instruments, is not a merely physical operation. It is physico-
physiologico-psychological, and the third component is the most
important because in every case it is in fact decisive.

**258.** The foregoing comments should throw added light on the
reason why the subject treated in these pages is called "Optics, the
science of vision" as opposed to "Optics, the science of images"
or seventeenth-century optics, arising from Kepler's hypothesis
about the telemetric triangle.

"To see an object" means that an effigy created by the mind is
put where the object is. This operation is almost always successful,
as is shown by the enormous practical utility of the sense of sight.
But "to see an image," in the same meaning of the words, may be
called an exceptional action. Moreover, it is very difficult to check,
because touch and the other senses do not perceive images, and
therefore cannot help us to confirm or deny the coincidence of the
effigy with the image. This question can be settled only by rational
analysis, as illustrated in the previous Chapters.

This fact readily explains why seventeenth-century optics could lead an untroubled life for over three centuries. Once faith in the basic hypothesis had been established and confirmed by very brilliant and highly significant achievements, it was deemed useless to *reason* with the aim of arriving at a possibly contrary conclusion.

The careful reader, who has undoubtedly noticed the expressions that I used in the preceding Chapters to avoid the term "image" altogether, now will easily understand my motive. Had I employed the phrase "to see an image" in the contexts where it seemed quite convenient and natural to do so, to that extent I would have admitted what I wanted to prove was wrong.

**259.** The concept of an image is very ancient, very difficult, and very interesting, and should be discussed with some thoroughness at this juncture.

In tracing the historical evolution of human thought about the mechanism of vision, we find that the earliest notions about the optical image go back to the *eidola* of the ancient Greek philosophers. These *eidola* were later transformed into the medieval *species*, which were *forms* as distinct from matter, or incorporeal appearances. Their predominant feature was their wholeness. The species of an apple was by its nature a whole, just as a material apple in the physical world is a whole. The idea that an apple could be a collection of points or elementary units, each acting by itself, encountered great difficulty in becoming rooted in the field of optics. From the time of its appearance in the writings of Ibn al-Haitham (Alhazen) to its perfection at the hands of Kepler about six centuries elapsed.

Kepler maintained that every point of an object emitted rays on its own account in all directions. At least some of these rays could be redirected so as to pass through another point. In each such point were concentrated the rays emitted by a single point of the object. Then the sum total of such points of concentration constituted the image of the object, point for point. The undisputed victory of Kepler's view led to the outright destruction of the *species*. Not only was the account of their formation rejected, but also the very conception of them as whole entities. Now the image of an apple was the aggregate of point images of the individual points of the material apple. One such image point, or a small group of them, could be considered separately, without thereby affecting the others at all.

**260.** This new outlook impelled optical studies to spurt forward, but it soon began to show signs of a crisis. For there are relatively few cases in which an optical system makes the waves emitted by a point of the object pass exactly through a point. Such accurate concentration is produced by a plane mirror and in a few other ways. In general the rays deviated by an optical system envelop a caustic. This fact was known before Kepler expounded his theory, and the knowledge deterred mathematicians devoted to logic and precision from entering upon what was destined to become the path of the future.

The new conception was initiated by Francesco Maurolico and given its complete and definitive form by Kepler. Both these men jumped to conclusions by what may be called a burst of genius and, above all, by stepping on strict logic, that is, by an approximation. The rule may be stated in more modern terms as follows. The rays envelop a caustic. We consider only the cusp of the caustic and ignore the rest of it. Our results will be closer to the truth, the smaller the angular aperture of the bundle of rays.

At first this rule was received with great enthusiasm, because it immediately brought order into questions that had baffled men's minds for centuries. Later, however, disappointment set in, since bundles of rays having too narrow an angular aperture do not give enough intensity. Everywhere the need was felt to widen the angular apertures. But then the rays obviously no longer passed through a point.

**261.** By logical standards the definition of the image was doomed. If the name "image" is bestowed on the point where the rays intersect, then obviously there is no image when the rays do not meet in a point. But experience showed that images were seen just the same. Hence by stepping on logic again and resorting to a kind of compromise, specialists in optics began to talk about *aberrations*. They said that when the rays pass through a point, the image is perfect. When the rays do not pass through a point, but *almost* do so, the image is imperfect. The displacements of the rays from the position that they must occupy for the image to be perfect are called aberrations. This is how students of the subject all over the world still think today, and nobody has ever remarked that what is lacking now is the definition of an image.

It is possible to talk about an image when the bundles of rays have a structure like a cone, though not exactly conical, because they pass, not through a point, but through a little neck. Whoever adopted this conception of an image would face the serious problem of defining the neck, that is, of ascertaining the dimensions of a disk such that when the rays pass inside its rim the image is present, but absent when they pass outside the rim. For if this boundary between presence and absence is not drawn, the disk may even be of enormous size. In other words, the image would *always* be present or would become something indefinite. The diameter of this disk has never been determined by anyone.

**262.** But the situation is worse than that. Consider an optical system having symmetry of revolution around an axis, and so constituted that the rays emitted by a point source on the axis emerge convergent and enveloping a caustic. Obviously, for reasons of symmetry, the caustic also must possess symmetry of revolution around the axis of the system. This is the very common case of simple lenses, spherical mirrors, etc.

All the rays striking a lens along a given *zone* (that is, along a circle with its center on the axis) emerge cutting the axis in a point. The aberration consists of the fact that as we pass from one zone to another of bigger radius, the point of intersection on the axis is displaced, generally approaching the lens.

In the simple case of Fig. 91 the point source is infinitely distant. Then the rays that pass through the lens in the paraxial zones, or zones very close to the axis, intersect at a point $F_p$, called the *paraxial focus*. The rays that pass through the lens along its rim or marginal zone cut the axis at a point $F_m$, known as the *marginal focus*. The rays that impinge upon any intermediate zone cross the axis at a point lying between $F_p$ and $F_m$, and form an absolutely regular cone.

Therefore every zone produces a perfect image of the source. Since the zones are infinite, the images also are infinite and constitute the segment $F_p F_m$.

Nobody has ever said so. Everybody says that the lens gives only one image, affected, however, by spherical aberration. But nobody

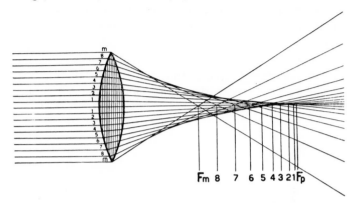

Fig. 91. Caustic formed by refraction through a converging lens

has yet said what he means by the image in this instance, and where he understands it to be located.

**263.** Because the radiation is refracted at different angles, there is another aspect of the behavior of lenses that is similar to the aspect discussed in § 262. If we consider a zone by itself, the rays corresponding to the various colors of the spectrum are deviated differently. Hence again there are infinite foci (in this instance, for the same zone) all distributed along the axis and corresponding to the several colors. Are there, for this reason too, infinite images of the same source, all produced by that particular zone of that particular lens? Nobody has ever said so. Everybody says that there is only one image, affected, however, by chromatic aberration. Yet nobody has said where he understands the image to be located.

Even more serious is the case of the so-called *astigmatic* optical system, such as a cylindrical or toric lens. An astigmatic lens has, not symmetry of revolution, but two planes of symmetry perpendicular to each other. Consider once more a point source $S$ (Fig. 92). The rays that it projects upon the lens are refracted so that they all fall on two segments perpendicular to each other. One of these two segments lies in one of the planes of symmetry, and the other in the other. The segments are called the *tangential focal line* ($F_1$) and the *sagittal focal line* ($F_2$), though ordinarily the word "line" is not used.

Ought we to say that such lenses give two images of the source? This is sometimes said, even though no vertices of cones are present; and there is talk of a sagittal image and a tangential image, all the more because they may even be infinitely distant from each other. But for the most part people continue to say that there is only one image, affected by astigmatism, and again they fail to indicate where it should be located.

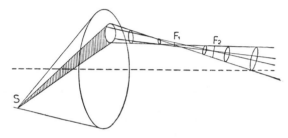

FIG. 92. Astigmatism of an oblique bundle of rays refracted by a converging lens

We are driven to the inescapable conclusion that what an image is and where it is are not known, the only thing known about it being that it is defective.

**264.** It is rather interesting that there is an optics of images and that it does not know how to define its fundamental entity. This curious fact may be added to the others previously adduced to show the artificiality of seventeenth-century optics and the necessity of replacing it by a structure agreeing better with experimental reality.

But now let us resume our analysis of how the concept of the image evolved. To follow the historical development, in these last few Sections I have been using the language of geometry, which is not altogether devoid of instructional value. Of course one of the weaknesses in the first definition of the image (the definition due to Kepler) was thàt it was formulated in terms that were too abstract. The rays were straight lines that perforce intersected in a point. In actual practice the point had to become a disk. This change was admitted for a long time, but without any precision. As was indicated in § 261, the disk has never been defined nor have its dimensions ever been determined.

The situation improved considerably with the advent of wave optics. Now the point source emits waves. These are deformed, becoming transformed into other waves which, if they are spherical, have a center. At this center a distribution occurs in the form of a centric.

The earlier abstractions, mentioned above, have been eliminated. The image of the source is no longer a point, but a centric of finite dimensions that are a function of $\lambda$ and of $D$, the diameter of the wave front. The tolerances met in practice begin to take shape, and when the waves are spherical, all goes well. In fact it is now possible to reverse the argument. If it can somehow be established that there is a centric at a wave's center of curvature, then it may be inferred that the wave producing the centric is spherical or at least that it does not differ from a spherical surface by quantities that are optically appreciable.

Thus there came to the forefront the concept of optical tolerance, which was later quantitatively defined by Lord Rayleigh as $\lambda/4$. Waves whose surfaces vary from a sphere of reference by less than $\lambda/4$ produce at their center diffraction figures that are still acceptable as centrics.

**265.** The progress made is considerable, but we are not yet at our goal. First of all, the question arises how we set about determining the form of the diffraction figure at the center of a wave in order to decide whether the figure is acceptable as a centric or not. In the second place, when the waves do not produce a centric, what is meant by the image?

To the second query we may reply at once that we do not say what the image is nor where it is; we say that it is present and that it is affected by aberration.

The usual answer to the first question seems simple and clear: take a look, with an eyepiece or a microscope. But as has already been pointed out in § 249, this response is a delusion. For "to look with a microscope" means, not to see what is in the object plane of the instrument, but only to convey the waves to the back of the observer's eye with no few and no trivial deformations along the way. All that will lead to the creation of an effigy, which may possibly be located by the observer in the object plane of the microscope and be regarded as what is found in that plane. But this whole operation is the doing of the observer. Then we have again bumped into the rules of the science of vision.

This outcome is inescapable. The reader will now be convinced that it is impossible to ask a question for the purpose of clarifying and defining the foundations of the optics of images without revealing that those foundations are modest and incomplete hypotheses intended as substitutes for the marvelous mechanism of vision, for which there is no substitute. Hence the reader will also have grasped the reason why, when any such question is raised, we are always thrown back on the rules of the science of vision.

**266.** Actually, on the basis of those rules, our concepts become much clearer and more satisfactory, even if more complex.

Suppose we want to look at what is happening in the plane that ought to contain the centers of the waves emerging from an optical system. We really have no means of doing so other than to make the waves proceed to the observer's retina. The effects that they produce there are then elaborated, depicted, projected outward, and attributed to the plane in which we are interested. But as a matter of fact we do not know what is in this plane, because we have no way of getting at it directly.

If we are to stay on solid ground and, so far as possible, exclude from consideration concepts that cannot be checked and confirmed by experience, we should not talk about an *aethereal image*, so to speak; that is, an image consisting of centers of aethereal waves and lacking all material support. For we can never say how it is made and what characteristics it has, without projecting it on a material body. In that case the only thing that we can ascertain is the *effects* of the waves on that body. But these effects are obviously influenced by the nature and properties of the body, and consequently differ from one instance to another.

The aethereal image is, therefore, something that cannot be handled experimentally, and accordingly should not be taken into account. These remarks may suffice to explain why it has not proved possible to define the image despite the fact that it has existed for three and a half centuries. More than one a priori definition on the model of a mathematical postulate can be given. But a definition corresponding to experience cannot be given of something that it is impossible to manipulate experimentally.

**267.** Hence if we want to reserve the term "image" for something that has experimental significance, our only resort is to give that name to the totality of the modifications impressed on a material surface. In other words, we must definitely avoid assuming any distribution of the radiant energy during its *supposed* propagation.

Critics of physics have pointed out that the propagation of the radiation is sheer hypothesis. Some have even concluded that the radiation is not propagated at all. According to them, this is the reason why centuries of effort have failed to find a mechanical model of the propagation, a model capable of accounting for all the phenomena known about the radiation. I myself decided in this book to adopt the wave model as serving the purpose better than its rivals. But I have no genuine faith in its real existence, nor in that of rays or photons.

Hence it is no wonder that in comparison with experience a definition has been found wanting that depends on the vertices of cones of nonexistent rays, or on the centers of nonexistent waves. But if we abandon this whole mass of hypotheses and refer to the modifications produced by the radiation on matter, we put ourselves on a much firmer footing, and then our concepts likewise can acquire a more precise content, because it corresponds to experimental reality.

For my part, therefore, I shall call a *distribution of energy effects on a layer of the retina* a *retinal image*.

**268.** This definition has the following five features that should be emphasized:

(*a*) it attributes the proper value to the layer whose participation as a detector is indispensable;

(*b*) it is independent of the model, whatever be its type, of the hypothetical propagation of the radiation;

(c) it agrees with the experimental finding that the image is always present whenever energy effects are distributed on the detecting layer;

(d) it is not subject to the optical system's condition, whether good or bad; and finally

(e) it assigns a definite position to the image by putting it on the detecting layer.

In the discussion that follows I shall try to make clear how practical this new definition of the image is and how closely it hews to everyday experience. But before doing so I must justify three terms used in the definition.

In the first place, I have adopted the word "image" for the sake of continuing a tradition and, above all, with the hope of persuading those who wish to enter this current of ideas to forget about the old aethereal or aerial optical image, call it what you will, just as the seventeenth century utilized the same name to designate the new aerial image that was displacing the older "species."

Secondly, I have qualified the image as "retinal" in order to indicate that the effect is obtained through the action of the radiation on the sensitive part of the eye.

Lastly, I have employed the expression "energy effects" to denote the modifications produced in the structure of the retina by the incident energy flux. I propose to show that the significant factor is, not the arriving energy, but what it does to the retinal layer.

This last remark contains the most important difference between the old conception and the new. The former view ran more or less like this. The image, whether perfect or affected by aberrations, is aerial or aethereal, whichever you want to call it, and it has a certain calculable structure. The effect it produces on the layer that it encounters, even at a distance of 1 $\mu$, is of no interest. If there is no effect, the detecting layer must be changed until one is found that feels what it is supposed to feel. On the other hand, the new view holds that the aethereal image is of no interest. Only what is produced in the detecting layer is of interest. This difference is no small matter, either philosophically or practically. We shall soon see its wide repercussions.

Having accepted the new definition, we may now say that the mind *generally* creates its effigies in conformity with the information reaching it from the retinal image.

**269.** This part of the discussion could be simplified by assuming the mind to be a passive and perfect agent, which assimilates all the information reaching it from the retina and depicts it with absolute fidelity. Then the effigy created would be the external projection of the retinal image. This would be true of course only from the geometrical point of view, because the photometric and colorimetric aspects always imply a profound mental elaboration. For the brightness corresponds approximately to the logarithm of the retinal irradiance; and the color is related, both in hue and saturation, to factors that are still very difficult to define.

We must not delude ourselves, however, about the perfection and passivity of the mind. Such an assumption would unquestionably simplify matters, but is of very doubtful validity.

**270.** The new definition of an image requires a clarification, which perhaps more than one reader has for some time been thinking of requesting. Much has been said here in disparagement of the aethereal images, which have been characterized as purely hypothetical conceptions, etc. But can they not be intercepted on a screen? Can they not be received on a photographic plate? Can they not be analyzed by a photoelectric system?

The objection is natural, but easily refuted. So far as possible we must free ourselves from the traditional concepts by returning to the point of view explained in §§ 21–23.

When radiation is projected upon a photoelectric cell and gives rise to electric currents or similar phenomena, we are entirely outside the realm of optics. To suppose that this method would make it possible to learn the structure of the aethereal optical image that activated the photoelectric system would be a delusion. What comes out at the end is a lot of information about the system that conveyed the radiation, and the system that detected it. This information can be evaluated most reasonably by extending to this detector the rules of the science of vision with appropriate modifications. It is particu-

larly interesting to see how closely photoelectric phenomena conform to these rules, which derive a new and important confirmation
from them.

**271.** When the radiation is projected upon a photosensitive
emulsion, it gives rise to chemical reactions. These generally lead
under later treatment to the deposit of a layer of granular metallic
silver, which looks opaque and dark when observed with the eye.
Once more we are entirely outside the realm of optics. To suppose
that this method would make it possible to learn the structure of the
aethereal optical image projected on the sensitive surface would be
another delusion. Again what we get at the end is a certain amount
of information about the action of the incident radiation on the
detecting system, later treated in a given manner. This information
too can be reasonably evaluated by extending to this detector the
rules of the science of vision with appropriate modifications. It is
remarkable how these rules lend themselves even better, if that is
possible, to the study of photographic phenomena than to the study
of visual phenomena. Anybody tracing the development of the
investigations that in the past twenty years have led to the outlook
expounded in these pages would discover that the most important
contribution has been made by observations and experiments in the
field of photography.

It is in fact extraordinarily difficult to take a photograph a
second time and have the two photographs so much alike that no
distinguishing characteristic can be found. It is indeed absolutely
impossible to do so, even in laboratory experiments, to such an
extent is the result affected by the detecting layer and its chemical
treatment.

We may say that photography too is a complex process, of which
the outcome depends on the following three factors: the radiant
energy emitted by the source, the distributing system, and the
detecting layer. In vision the three corresponding factors are the
radiant energy, the distributing system, and the detecting retina.
Hence, just as we defined the retinal image, we may define the
photographic image as a "distribution of energy effects on a photosensitive emulsion." We shall soon (§§ 285–286) have the oppor-

tunity to point out the practical utility of this definition of the photographic image.

**272.** Now we must take up the case of the figures that are seen on a screen when, to use a conventional expression, a "real image" is projected upon it.

If we examine this phenomenon without any theoretical preconceptions, we see the following situation. The source emits waves, which pass through a converging optical system and are made convergent, becoming more or less spherical with their centers in a certain plane. Then this plane contains the locus of the centers of the waves emerging from the optical system. In this plane put a screen, that is, a diffusing surface, which acts either by reflection or transmission. This means that when the waves arrive, they cause the material they strike to emit new waves, generally spherical, in all directions or in a half-space, the intensity being predominant in certain directions rather than in others.

These diffused waves may be received by the eyes of an observer. On the basis of the information coming in from the retinas of his eyes, his mind creates an effigy that is located, if all goes well, on the screen.

Therefore the figure that is seen is again an effigy created by the mind of the observer. The function of the screen is to change the course of the waves in such a way that the information on the retinas of the observer is modified. Theoretically the screen's function does not differ from that of an eyepiece or a microscope, with which some observers imagine they see what happens in the plane of the centers of the waves.

**273.** The change that the screen produces in the course of the waves is far more profound than that produced by an eyepiece, and not always more advantageous in practice. The case of the screen somewhat resembles that of the photographic plate. The waves absorbed by the material of the screen are re-emitted in the form of waves; those absorbed by a photographic emulsion are transformed by chemical processes. The behavior of the diffusing screen, therefore, by reason of its analogy with both the microscope and the photographic plate, adds nothing essentially new to our under-standing of the problem.

If the screen holds any interest for us, it does so because it aided in the establishment of seventeenth-century optics. As was pointed out in § 46, Kepler realized that when the radiation reaches the eyes by way of an optical system, the placement of the effigies by the mind is too subjective and hence too arbitrary to provide a basis for enunciating laws of physics. Hence he thought of reducing to a minimum the intervention of so capricious an operator by limiting the activity of the eye to the examination of figures on screens, while preventing its receiving directly the waves emerging from the optical system. He called the figures on the screens "pictures," and applied the term "images of things" to the figures that the observer would have seen in the absence of a screen by receiving in his eyes the waves emerging from the optical system.

Even though this distinction had no success terminologically, as was indicated in § 51, it nevertheless exerted a highly important effect from the scientific point of view. For it contributed mightily to pushing the study of optics down the slope to which it had been guided by seventeenth-century philosophy, that slope of supposed objectivity according to which it was necessary so far as possible to eliminate the intervention of the eyes and, above all, of the mind.

**274.** If Kepler's proposal is scrutinized in the light of the ideas set forth here, it takes on the following meaning. When the eyes receive waves emerging from an optical system, they send to the mind strange reports, which in general are hard to reconcile with the information furnished by the environment and the memory. Most of this information is supplied by the experience of direct vision, a vast experience extending from minute to minute over years and years, and therefore capable of conferring a very high degree of probability on the results of vision when they are checked by the other senses. In the face of disagreement between optical reports and information derived from the memory and the environ-ment, the mind usually prefers the latter, with consequences that were discussed in the preceding Chapters.

When a screen is put in the path of the waves, those that reach the eyes have their centers in points on the screen. Therefore the eye operates by direct vision, and the whole conflict that embarrasses the mind is eliminated. The information based on the memory and the environment is set aside or, rather, is replaced by much simpler information.

**275.** Merely a recording function is thereby left for the eye. Co-operation between the optical system and the ocular system is completely removed. In relation to the energy distribution on the screen, the action of the eye may be replaced by that of another detector of energy, such as a photographic plate or a photoelectric analysis.

Suppose we take the following setup: an object that emits waves; a converging optical system that receives the divergent waves and makes them convergent; a screen that intercepts these waves; and, finally, an observer whose eyes are stimulated by the waves re-emitted by the screen. This whole process may be divided into two phases. The first consists of the emission, deformation, and absorption of the waves; the second phase includes the emission of waves by the screen, the reception of them by the eyes, and vision by the observer.

The first phase has no more to do with optics than a photograph has. Therefore, when we use a screen to perform our experiments, we act as we do when we take a photograph and then look at it. There is no essential difference between the two operations. Hence, just as we spoke in § 271 of a "photographic image," we may now speak of an "image on a screen." Just as the former means a "distribution of energy effects on a photosensitive emulsion," the latter must mean a "distribution of energy effects on a diffusing substance." And just as the former is affected by the nature of the emulsion and its chemical treatment, so the latter is affected by the structure of the diffusing surface.

**276.** These basic concepts find their best confirmation in the examination of a problem that is apparently technical but actually thoroughly scientific, namely, the definition of a *good* image (in the conventional sense of that term).

By what criteria should this "goodness" be defined? This is an extremely difficult question, and in many cases it is still wide open.

Geometrical optics has shown itself utterly powerless to cope with it. Wave optics has yielded interesting results, but in the end has had to forego a complete solution of it.

Geometrical optics gave the name "image of a point object" to the point through which all the rays emerging from an optical system were supposed to pass, and thereby laid claim to mathematical perfection. If the rays did not all happen to pass through a point, the students of the subject limited themselves to saying that there were aberrations.

Since mathematical perfection is naturally unattainable in practice, all images had to be considered as affected by aberrations. However, there were degrees of aberration. Then which were the good images, and which not? That question was answered by nobody. The aforementioned students entrenched themselves behind a dignified disdain for such a problem. It was a purely technical matter.

Thus when they had demonstrated that a plane mirror gave perfect images, their task was finished. To the objection that no material mirror was ever *plane*, like a mathematical plane, they replied that this concerned, not them, but the masters of the art of making plane mirrors.

In essence, geometrical optics confined itself to saying only this: the image of a point source is better, the closer the emerging rays pass to the point image. This statement is not entirely correct, as was shown by wave optics.

**277.** According to the wave theory, as we saw in § 264, the center of a wave emitted by a point source and modified by an optical system is a centric, whose angular dimensions depend on $\lambda$ and on $D$, the diameter of the diaphragm that limits the wave front.

The perfect image is, therefore, a centric. Since this figure can be obtained, it is produced, not by ideally perfect waves, but by actual waves. In other words, a centric can be formed even by waves which, like those always met in practice, are not completely spherical, provided that the departures from a sphere are kept within proper limits. This time those limits have been defined by Lord Rayleigh's well-known "rule of a quarter of a wave."

This is a considerable step forward. But let us not make the mistake of attributing a theoretical foundation to it, as is sometimes done. For according to the wave theory, when the convergent wave under consideration is not spherical, the energy distribution around its center of curvature is not a standard centric, but something different. Of course the differences may be small, if the deformations of the wave also are small. When these deformations are less than $\lambda/4$, the differences from a centric may be considered negligible, as has been shown experimentally.

How this was shown was discussed in § 245 and will now be treated more thoroughly in § 279. But first I wish to point out that even the wave theory was unable to say anything when the waves show deformations exceeding $\lambda/4$. Is the image good or bad? If it is not good, what are the effects of its imperfections? Only vague and inconclusive answers have been given to these questions. No numerical determination has been proposed.

**278**. The situation looks entirely different when viewed from the standpoint of the science of vision or, more generally, of *energy optics*.

If we bear in mind that there is a receiver-detector of the radiation, we immediately draw the conclusion that the detector is the final judge of the goodness of what it receives. For if there are two hypothetically different trains of waves that nevertheless produce the same effect on the receiver, they must be deemed equal.

The objection may be raised that hypothetically different trains of waves ought to produce different effects on the receiver. This is true in mathematics, but not in practice. For in all experimental matters there is a tolerance. When the hypothetical differences are such that the differences in the effects fall within the tolerance under consideration, it is as though the hypothetical differences did not exist. The decisive rule in this regard has already been stated many times. Two agents are said to be different when the effects produced by them on the receiver are perceptibly different.

Tolerance is a characteristic of the receiver, and may differ widely from one to another. There are no grounds for assuming a general tolerance. At the very most we may seek a *statistical value* for a group of receivers with similar characteristics, if such a value exists.

For an image to be good or bad now means that its defects, which are always present, fall within or without the tolerance of the receiver. Hence it follows at once that "goodness" defined in the absolute sense has no meaning. On the other hand, we may speak of goodness for a certain receiver.

**279.** This analysis explains at once why both geometrical optics and wave optics have been utterly unable to solve the problem of "goodness." In this connection they paid no attention to the receiver, which is the only factor capable of supplying a standard to decide the matter.

Moreover, the sole contribution of wave theory to this question is the "rule of a quarter of a wave," which was discovered experimentally by resorting to certain observations made with the eye. This rule provides a statistical value that is significant for vision with the eye.

But the value may possibly be regarded as having theoretical significance too and as being a final answer. It is highly probable that in practice we ultimately have to resort to determining one or more statistical values. But we must not let the prevailing philosophical current pull us into exaggerations that present matters falsely.

The desire at all costs to eliminate the receiver from consideration has led to the following line of conduct. Once some experiments (performed with the eye, of course) had determined a tolerance for the perception of deformations in the centric, a way was found to express that tolerance in terms of the deformation of the wave. This way led to the enunciation of the "rule of a quarter of a wave," which considers deformations of a wave less than $\lambda/4$ as negligible. Then it was completely forgotten that this rule was the result of some measurements made with a given eye or a few eyes. Everything possible was done to have this origin forgotten. The rule of a quarter of a wave thus became a rule unto itself, as if it were based on principles and proofs independent of the observer.

Hence the natural deduction was that when an optical system deformed the waves by giving them a plane form, or a spherical form with crests and troughs less than $\lambda/4$, it was to be considered perfect in itself, independently of the observer.

This deduction, at bottom, amounts to saying that the tolerance established with those few measurements and with those few eyes is always true for all eyes and for all experimental conditions. This statement obviously is not true.

**280.** For not only is the behavior of an eye highly individual, varying from one eye to another even of the same observer, but it is also a function of not a few variables, the most important of which is the energy reaching the retina.

To cope with variations in behavior, we may resort to statistical values by grouping eyes in categories according to some criterion. We may, for example, measure the visual acuity (or the resolving

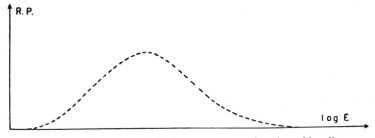

Fig. 93. Resolving power of the human eye as a function of irradiance

power of the eye in direct vision) in terms of the standards adopted by the ophthalmologists, and use the Snellen notation in determining the tolerances. As our basis we may say that the rule of a quarter of a wave indicates the tolerance for eyes with 20/20 vision.

But far more serious and more interesting is the fact that the resolving power of an eye is not constant because, as we saw in § 221, it depends essentially on the irradiance of the retina. As a function of this irradiance, it shows a central maximum with descending slopes for both increasing and decreasing irradiances. The curve of this function (Fig. 93) is one of the fundamental parts of energy optics.

As is clear from the theory set forth in § 223, the position of the maximum is in its turn a function of the sensitivity of the detecting layer or retina. This sensitivity in turn varies as a function of numerous factors, first among which is adaptation, then the age of the observer, his diet, state of fatigue, etc.

Moreover, the value of the maximum is affected most of all by the contrast between the various portions of the irradiance at the back of the eye. Hence this value is influenced by the disturbances due to the diffusion of the radiation in the retina and humors of the eye, and by the presence of intense sources in the field, etc.

Therefore the tolerance is a quantity that is affected more or less indirectly by all these factors.

**281.** Hence from the practical point of view the problem is tremendously complicated and becomes increasingly indefinite and hazy. More than one reader, especially if interested in technical optics, may at this point refuse to enter so muddled a field, and prefer to return to the schematic and precise simplicity of the older way of viewing the question. But the older rules were limited in range. They boiled down in substance to the rule of a quarter of a wave, and supplied no data outside the zone of optical perfection. Furthermore, I shall cite several typical cases in which the older rules, though simple and stimulating, led to real errors not without serious consequences.

To obtain a telescope capable of yielding emergent waves deformed by less than a quarter of a wave, several lenses must be combined to correct the chromatic and spherical aberrations of the individual components. But every lens involves a loss of energy flux through reflection, absorption, etc. All this is of slight importance when the flux at our disposal is sizable, and the loss due to the interposition of the instrument in the path of the radiation is either negligible or easily compensated by the adaptation of the observer's eye.

But the importance is not slight when we must operate in conditions of feeble radiation. To lose a percentage of it may mean to fall below the threshold of sensitivity, or at any rate to drop into regions of very low resolving power, namely, those at the left of the curve in Fig. 93.

If we go below the threshold, the observer sees nothing. But instead of seeking to correct as far as a quarter of a wave, we may use an instrument (of the same dimensions, of course) consisting of two very simple lenses, one for the objective and the other for the eyepiece (a telescope cannot be made with a smaller number of lenses). Then the loss of energy flux would not be such as to bring us below the threshold, and the observer would see something. That is always better than seeing nothing.

**282.** Let us apply this style of reasoning to an important instrument like a range-finder of the coincidence type, which necessarily has a very low energy output. It is called upon to do its work, let us remember, above all under conditions of comparatively feeble radiation, like dawn and twilight, for obvious tactical reasons. Installing several lenses and plates to obtain an extremely fine optical correction was an excellent policy for range-finders designed to function in full daylight. But we must conclude that not infrequently it was a disastrous policy for range-finders actually employed at the threshold of visibility.

If today a range-finder were made slightly inaccurate for the purpose of maximizing its light, it would certainly be rejected by the official inspectors, who would doubtless choose a very bright place to examine with precision the deformations of the emerging waves. Such conduct, heretofore considered obligatory, involves a definite error according to the new theory. I could continue along this line at some length. The definition of the energy conditions under which an instrument is required to function is an essential consideration in giving it the most rational structure.

**283.** Furthermore, there are observers with less than 20/20 vision. Obviously they do not need so fine a tolerance as a quarter-wave. Yet no special attention has ever been paid to them. Nobody has ever thought that many instruments showing defects intolerable for observers with very acute sight may be ideal for observers with inferior vision.

If particular consideration were given to these observers, who are far from rare, rules might be discovered for designing and manufacturing cheaper and more useful instruments. For instance, observers with poor vision probably would derive greater benefit from instruments with extreme magnification even though that entailed a greater distortion of the field.

We are entering upon a new road, which heretofore has been closed and concealed by the philosophical preconceptions of seventeenth-century optics and by an exaggerated desire to schematize phenomena whose nature is very complex.

**284.** To avoid going too far into questions that are assuming a distinctly technical character, let us turn back to consider the figures intercepted on a photographic plate or on a screen. We called these figures "photographic images" (§ 271) and "images on a screen" (§ 275). The analysis to be made applies equally to both, and therefore we shall deal only with the latter.

Let the points of a linear object project spherical waves upon an optical system, which makes the waves convergent but still sensibly

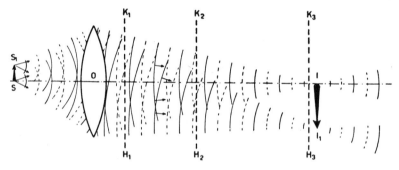

Fig. 94. Waves emerging from a converging lens and intercepted on a screen at various distances from the lens

spherical. To simplify the illustration, Fig. 94 shows only two trains of waves with their centers at $S$ and $S_1$. The lens $O$ makes the waves converge at the points $I$ and $I_1$. Between these two points lie countless other centers, so that the locus of the centers of the waves emerging from the lens extends from $I$ to $I_1$.

Intercept these waves on a screen. Place it in a random position, and then move it closer to or farther from the lens. What is observed during this movement?

When the screen is very close to the lens, as at $H_1K_1$, all the waves strike the same area of the screen, and therefore the irradiance is uniform. As the screen is moved away from the lens to a position like $H_2K_2$, differences start to show up. The irradiated area contracts toward one direction and hence acquires an elongated form, whose edges are hazy. But if the points $S$ and $S_1$ of the object emit waves of different intensity or of different $\lambda$, darker and brighter regions begin to appear as well as regions in which a hue of moderate saturation commences to be visible. Now move the screen farther from the lens, say to $H_3K_3$, where it is closer to the locus of the centers of the waves. The illuminated area lengthens and contracts; the brighter regions are more clearly marked off from the darker; the regions of one hue contrast more sharply with those of another; and the saturations tend toward the maximum. As the screen continues to recede from the lens, the reverse process is observed. The illuminated area expands, showing increasing uniformity in brightness and hue, while the saturation decreases continuously.

**285.** What is done in such cases is well known. The screen is moved back and forth until it shows the details of the figure with maximum sharpness, maximum brightness, and maximum saturation (if there are any colors). The screen is stopped in this position, which is therefore defined by an estimate of these maxima.

This estimate obviously depends on the means of observation. If we observe with the naked eye, we usually find a sizable interval within which the screen may be moved without any appreciable diminution of the characteristics observed at their maximum. The size of the interval increases with a decrease in the angular aperture of the waves whose centers are $I \ldots I_1$, with a weakening of the irradiance on the screen, and with a lowering of the resolving power of the observer's eye or eyes.

Those who wish to place the screen in the position of maximum effectiveness with the greatest possible precision resort to observation with an eyepiece or a microscope. This practice is particularly common in determining the setting of photographic plates in cameras.

Let us examine the significance of this operation, which is apparently so commonplace. According to our conceptions, it is really a simple matter. The image on the screen (or on the photographic plate) is always present, however far it may be from the lens (provided of course that it is perceptible). But this statement refers to the image as it has been defined here, namely, as a distribution of energy effects. The difference between these effects varies in the images appearing at various positions of the screen. The position or interval in which this differentiation is a maximum may be determined by recourse to more or less precise means of observation.

The naked eye is one of these means. It receives the waves emitted by the irradiated elements of the screen, and projects these waves upon the retina to have them analyzed by the mind. The eye equipped with a microscope acts in the same way, although with greater accuracy if the available energy so permits. But systems of another kind, photographic or photoelectric, may also be devised.

**286.** There is, then, an interval of space such that when the screen is placed within it, the energy effects on the screen show the maximum differentiation or what is commonly known as the maximum sharpness of detail. This interval decreases with an increase in the resolving power of the means of inspection.

According to the accepted view, what is seen on the screen ought to be the so-called *real image*, that is, the locus of the centers of the waves emerging from the lens or, if you please, the aggregate of the centrics corresponding to the individual points of the object. What is visible on the screen when the latter does not coincide with the line $II_1$ is termed the "*extrafocal image*," the label "*sharp image*" being reserved for the ideal position.

This name "extrafocal image" is a masterpiece. If the image of a point is a centric, the extrafocal image that appears when there is no centric on the screen is not an image. Therefore an extrafocal image means an *image that is not an image*.

The expression "image out of focus" has in fact no official standing. It was born in practice, especially the practice of photography. It was coined by photographers who, when they failed to put the plate in the interval of maximum differentiation, still obtained a photograph, but one inferior to what they desired. As a photograph actually taken and developed, it presented an image. By the characteristics of its details it showed that it had been taken outside the interval of maximum differentiation or with the plate too far from the focal plane (for infinitely distant objects) and was therefore out of focus.

This elementary and practical reasoning jibes perfectly with the concepts of the new energy optics. From the standpoint of classical wave optics it simply leads to a contradiction in terms.

**287.** But this is not the whole story. Study of the diffraction of waves demonstrates that a centric is formed not only in the plane that passes through the center of a spherical wave at right angles to its axis, but also a little in front and a little in back of this plane. There is, then, a *depth of focus*. This conclusion too is without doubt a step ahead of the geometrical theory, for which tolerances do not exist. But the step is not yet big enough. For depth of focus as thus defined is a magnitude characteristic of a wave. It is a function only of $\lambda$ and of the angular aperture of the wave. Therefore it is independent of the observer. As we well know, however, experience shows that things go differently.

Placing a screen near a "real image," understood in the classical sense as the locus of the centers of the waves made *convergent* by an optical system, cannot pretend to be done with the objectivity and certainty generally attributed to this process. Nevertheless, the uncertainty involved is a mere trifle as compared to the unreliability of the locations of the effigies created when the same waves are received directly by the eye. Consequently Kepler's suggestion was really invaluable when he advised us to consider the "pictures" and not the "images of things."

Hence the term "real image," interpreted in the classical way, may still be useful to denote the "locus of the centers of curvature of the waves made convergent" by an optical system. To be sure, the latter expression is long, but it is much clearer and more realistic. For it does not in the least imply that a visual process must be connected with such a locus of centers. On the other hand, when people talk about "images," it is always taken for granted that they are discussing something *seen*.

**288.** What deserves to be banished from *general* treatments of optics is the *virtual image*. This should be understood as the locus of the centers of curvature of waves that emerge divergent from an optical system. Such waves never give rise to "images on a screen" or "photographic images" or "photoelectric images," interpreted in the new manner. The only way to utilize these waves is to send them into an eye, so that the mechanism of vision may be activated and the ensuing effigies created. In very few cases are these effigies placed near the virtual images, as was shown by the extended and minute analysis in Chapter IV of vision by means of optical systems.

The creation and diffusion of the widespread belief that what is seen with an optical instrument is the virtual image it produces constitute one of the biggest mistakes of classical optics. It is really amazing that so obviously false a proposition should be looked upon with complete unanimity as an indisputable truth.

The virtual image is purely a mathematical fiction. It may be useful as an intermediate solution in the study of complex optical systems. But its application should be limited to those who specialize in performing optical calculations or designing optical instruments. Specialists, as such, may be expected to have a proper understanding of the concept's significance, and to realize the fact that hardly ever does a figure seen by an observer correspond to a virtual image.

Were the definition and discussion of the virtual image eliminated from the general treatment of optics in all schools, not only would many students be happier, but many wrong ideas would be prevented from entering the intellectual background of broad sections of the public.

**289.** The description of the long and complex process that culminates in vision winds up by being rather disconcerting.

We have considered a group of atoms which, by reason of their own motions or those of their components, emit aethereal waves. These reach an eye or a pair of eyes, enter within, and end up in a bunch of cells. There they are absorbed and cease to be waves. As a result the cells undergo modifications and transmit impulses along the optic nerves to the brain. Then the impulses are turned over to the mind, which studies their characteristics and compares these with the mass of information in its files. In conclusion it creates a luminous and colored figure, which it places where it believes the initial group of atoms to be.

Assume that everything has worked out perfectly and that the effigy occupies precisely the same position in space as that group. Shall we say that the effigy and the group of atoms are equal, or that they are identical? Obviously not.

The effigy is bright and colored, continuous and motionless. The group of atoms is discontinuous and its individual components are in motion; in reference to them, the descriptions "bright and colored" have no meaning.

Therefore, if the group of atoms is called the object, we cannot say that the effigy is the object. But we may say that it is a result of the object, a result attained by a mental operation based on the signals emitted by the object. Hence those who assert that they *see the object* are very optimistic about the correspondence between appearance and reality.

**290.** There are also those who are highly optimistic about other ways of knowing the object, for example, by means of a photograph. There are those who even swear to the absolute objectivity of the photograph and to its faithfulness, since there is nothing subjective about it.

Let us take a good look at what this objectivity and faithfulness consist of. The atoms of the object emit their waves. These pass through other blocks of atoms, whose form and properties are such that they let the waves pass, but deform them, make them convergent, and convey them to a gelatinous layer containing crystalline clumps formed by ions of silver and bromine. The waves reach some of these clumps, are absorbed by them, and therefore produce modifications in them.

These modified and nonmodified clumps are subjected to certain material agents which, after contact and an interchange of ions, finally leave grains of metallic silver where the clumps of bromine and silver were modified, whereas the clumps that did not undergo any modification are dissolved and carried away.

Thus there remains a gelatinous layer in which grains of silver are immersed. Between the distribution of these grains and the atoms of the object there is a certain correspondence. May we say that when we look at the gelatinous layer obtained in this way, we see the object? It would take a confirmed optimist to say anything of the sort. May we say that when we look at the layer, we can know how the object is made? It would take very much of an optimist to say this too.

The only thing that can be said is that the distribution of the grains of silver in the gelatine is a result of the emission of waves by the object, a result attained by a chemical operation.

**291.** But in that case what can we do to know how the object is made? This is a very old question. Anybody who carefully scrutinizes the means that our ego and our science have at their disposal for the purpose of investigating the structure of the external world is still obliged to refrain from answering the question.

We must still have deep faith in the means at our disposal for us to say that with their aid we can dig out much information about the structure of the external world. We can learn much, but not everything. For everything is so much that, as far as our means are concerned, it may be called infinite. Therefore every representation that we make for ourselves of the external world will always be perforce imperfect and partial, if not false.

So pessimistic a conclusion would be harmful if it were communicated to students. It was one of the blemishes of medieval philosophy. Strictly logical reasoning of this type deterred philosophers and researchers from studying lenses in the full three centuries intervening between the invention of eyeglasses and the invention of the telescope.

It took a declaration of faith by a man of faith like Galileo to change the course of events and initiate the magnificent scientific development that followed the invention of the telescope.

The men of flawless logic said that the information that we can have about the structure of the external world will never be complete; therefore, being partial, it is false.

Galileo said that the information we can have about the structure of the external world will never be complete; but it is information and therefore, even if partial, it is true.

While his adversaries did not want to look into the telescope to make sure that they would not be deceived, he staked his life and his future on the truth of what he saw by means of the telescope.

**292.** Today Galileo's faith, without any opposition, dominates the environment in which experimental science is carried on. Its practitioners have perhaps exaggerated, however, and a redressing of the balance may be useful for the sake of avoiding errors and illusions.

This is what I have attempted to do in this book. But I would not in the least want its effect to go so far as to destroy in those who have had the patience to read it to the end their faith in observation and science. For a logic that is irreproachable and infallible, but negative and sterile, is far less desirable than a scientific faith which, even if deceptive, is active and productive.

# APPENDIX

*Elements of Wave Motion*

# CHAPTER VII

# Notes on the Kinematics and Dynamics of Oscillatory Motions

◇◇◇◇◇◇◇◇◇◇◇◇

**293.** In the preceding Chapters we have often used rules about the propagation of waves, especially through a medium that presented anomalies or discontinuities. Not a few of these rules originated in the field of optics, and for a long time the study of them was known as "optics." Some people still call this study *physical optics;* others, *wave optics.* Actually it is purely a branch of mechanics, and its conclusions are valid for any type of wave, whether material or aethereal. It is, therefore, entirely improper to speak of optics in this connection, because we could with equal right speak of acoustics, electromagnetism, or the theory of earthquakes.

Our subject, then, is "wave motion." We may talk about other sciences when we apply the conclusions of this subject to particular waves and consider special receivers, such as the eye, ear, radio receiver, photographic emulsion, etc., as we did in the preceding Chapters.

For the convenience of the reader, I have assembled in the following Sections those ideas about wave motion that were used in the science of vision.

**294.** When the motion of a point $P$ along a straight line is described by the equation

$$s = A \sin \left( 2\pi \, \frac{t}{T} - \varphi \right)$$

the point $P$ is said to execute a *simple harmonic motion*. In the equation $s$ stands for the *displacement* or the distance of the point $P$ from the starting point $O$; $t$ is the time, an independent variable; $A$, $T$ and $\varphi$ are constants, whose meaning will now be determined.

Whatever the values of $t$, $T$, and $\varphi$ may be, $s$ can never exceed $A$ in absolute value. Once the values of $T$ and $\varphi$ have been fixed, as $t$ varies within a long enough interval, $s$ assumes the values $+A$ and $-A$ alternatively, and in succession all the values included between these two extremes. This means that the point $P$ performs a back and forth motion.

The constant $A$, which thus indicates the maximum value of $s$ in absolute terms and *half* of the whole path of the point along the trajectory, is called the *amplitude*.

**295.** The displacement $s$ can assume the same value for different values of the time $t$. Thus let $t_1$ and $t_2$ be two different instants. The two corresponding displacements $s_1$ and $s_2$ are given by

$$s_1 = A \sin\left(\frac{2\pi}{T} t_1 - \varphi\right); \qquad s_2 = A \sin\left(\frac{2\pi}{T} t_2 - \varphi\right)$$

It is possible to have $s_1 = s_2$, provided that

$$\sin\left(\frac{2\pi}{T} t_1 - \varphi\right) = \sin\left(\frac{2\pi}{T} t_2 - \varphi\right)$$

This condition is satisfied whenever

$$\frac{2\pi}{T} t_2 - \varphi = \frac{2\pi}{T} t_1 - \varphi \pm 2k\pi$$

where $k$ is any integer. Hence it follows that

$$t_2 - t_1 = \pm kT$$

Therefore whenever the time varies by $T$, $s$ assumes the same value.

If we calculate the velocity of the point $P$

$$v = \frac{ds}{dt} = \frac{2\pi A}{T} \cos\left(\frac{2\pi}{T} t - \varphi\right)$$

we see at once that the velocity is equal at instants $t_1$ and $t_2$ if

$$t_2 - t_1 = \pm kT$$

The same value is assumed by $s$ also when $t_1$ and $t_2$ are connected by the relation

$$\frac{2\pi}{T} t_2 - \varphi = \pi - \left(\frac{2\pi}{T} t_1 - \varphi\right) \pm 2k\pi$$

But in this case the velocity at instant $t_2$ is equal and contrary to the velocity at instant $t_1$.

This interval of time $T$, within which the moving point returns to the same position with velocity equal in magnitude and sign (this amounts to saying that the identical motion is repeated), is called the *period*. Simple harmonic motion, then, is a *periodic motion*.

The reciprocal $1/T$ of the period is called the *frequency*. It gives the number of oscillations completed in a unit of time by the moving point. When $t = 0$, $s = -A \sin \varphi$.

**296.** Two harmonic motions that are equal in amplitude and in period may still be different because they have different values of the last constant $\varphi$; then

$$s_1 = A \sin\left(\frac{2\pi}{T} t - \varphi_1\right) \qquad s_2 = A \sin\left(\frac{2\pi}{T} t - \varphi_2\right)$$

Given a value $s_1$ at an instant $t_1$

$$s_1 = A \sin\left(\frac{2\pi}{T} t_1 - \varphi_1\right)$$

it is possible to find an equal value $s_2$ at another instant $t_2$, provided that

$$s_2 = A \sin\left(\frac{2\pi}{T} t_2 - \varphi_2\right) = s_1 = A \sin\left(\frac{2\pi}{T} t_1 - \varphi_1\right)$$

Of all the instants $t_1$ and $t_2$ to which this equation applies, those are closest to each other that satisfy the condition

$$\frac{2\pi}{T} t_2 - \varphi_2 = \frac{2\pi}{T} t_1 - \varphi_1$$

This may also be written

$$\varphi_2 - \varphi_1 = \frac{2\pi}{T} (t_2 - t_1); \qquad t_2 - t_1 = \frac{T}{2\pi} (\varphi_2 - \varphi_1) = \text{a constant}$$

In other words, if $P_1$ and $P_2$ are the points subject to the displacements $s_1$ and $s_2$, respectively, the former may be said to execute a motion identical with the latter's but displaced in time with respect to it. Thus if $\varphi_2 > \varphi_1$, $P_1$ always leads $P_2$ by a constant interval of time, which is given by the last equation above. The constant $\varphi$ is called the *phase*, and the expression $\varphi_2 - \varphi_1$ is termed the *phase difference*. If $\varphi_1 = \varphi_2$, the two motions are said to be *in phase*. If $\varphi = 0$, for $t = 0$, $s = 0$.

The harmonic motion given by the simplest equation

$$s = A \sin \frac{2\pi}{T} t$$

differs from the motion described by the most general equation

$$s = A \sin \left( \frac{2\pi}{T} t - \varphi \right)$$

only because at the instant $t = 0$ the moving point is at the starting point $O$ in the middle of its trajectory, and returns to it after every period $t = kT$, $k$ being any integer.

The point moving in accordance with the general equation therefore executes a motion that is the same as that of the other point, but *follows* it by a constant time $\tau$, given by

$$\tau = \frac{T}{2\pi} \varphi$$

Then since

$$\varphi = \frac{2\pi}{T} \tau$$

the general equation for simple harmonic motion may also be written

$$s = A \sin \frac{2\pi}{T} (t - \tau)$$

Since

$$\frac{\varphi}{2\pi} = \frac{\tau}{T}$$

what was said about $\varphi$ may be applied to $\tau$. For any value $\tau = kT$, where $k$ is a positive or negative integer, $s$ has the same value as for

$\tau = 0$. When $\tau = \pm \dfrac{T}{2}$, that is, when $\varphi = \pm \pi$, the two motions differ. They are

$$s_1 = A \sin \frac{2\pi}{T} t \quad \text{and} \quad s_2 = A \sin \frac{2\pi}{T} \left( t + \frac{T}{2} \right) = A \sin \left( \frac{2\pi t}{T} \pm \pi \right)$$

Since the second may be written

$$s_2 = -A \sin \frac{2\pi}{T} t$$

we see at once that $s_2$ is equal to the motion $s_1$, but in the opposite direction.

Two motions, like those just considered, one of which follows the other by a half-period, or that are out of phase by a half-period, are said to be *in opposite phase* or *in opposition*. When the phase difference between two motions is $\pm \dfrac{\pi}{2}$, or one follows the other by $\pm \dfrac{T}{4}$, the two motions are said to be *in quadrature*.

**297.** To represent harmonic motions graphically, the values of $s$ are indicated as ordinates, the values of $t$ as abscissas, on two rec-

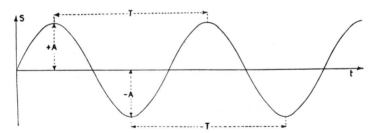

Fig. 95. Wave in a vibrating string

tangular axes $(s, t)$. In this way a sinusoid is obtained, like that in Fig. 95. Fig. 96 shows two motions equal in period and in phase, but differing in amplitude; Fig. 97, two motions differing only in period; Fig. 98, two motions in opposite phase; and Fig. 99, two motions in quadrature.

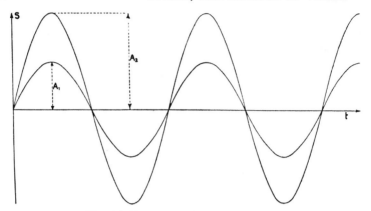

Fig. 96. Two waves of different amplitude

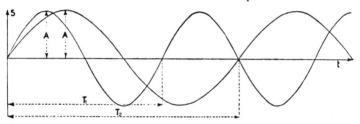

Fig. 97. Two waves of different wave length

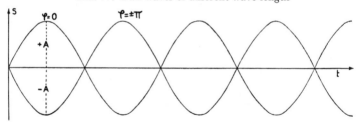

Fig. 98. Two equal waves in opposite phase

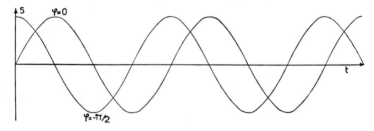

Fig. 99. Two equal waves in quadrature

**298.** Let a point $P$ be subject at the same time to $n$ simple harmonic motions, whose trajectories all coincide with the same straight line. Let the motions have the starting point $O$ in common, and let them be described by the equations

$$s_1 = A_1 \sin \left( \frac{2\pi}{T_1} t - \varphi_1 \right)$$

$$s_2 = A_2 \sin \left( \frac{2\pi}{T_2} t - \varphi_2 \right)$$

$$\cdot \quad \cdot \quad \cdot \quad \cdot \quad \cdot \quad \cdot \quad \cdot \quad \cdot \quad \cdot \quad \cdot$$

$$s_n = A_n \sin \left( \frac{2\pi}{T_n} t - \varphi_n \right)$$

The resultant displacement $s$ at instant $t$ is the algebraic sum of the separate displacements

$$s = s_1 + s_2 + \cdots + s_n = \sum_1^n {}_i A_i \sin \left( \frac{2\pi}{T_i} t - \varphi_i \right)$$

This operation may also be carried out graphically. On the same pair of Cartesian coordinates, draw the sinusoids corresponding to the separate equations (Fig. 100). Then trace the curve which, for

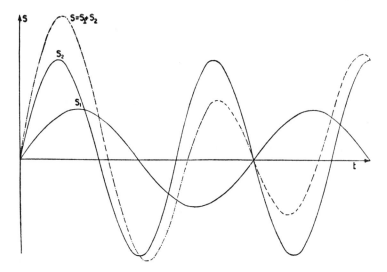

FIG. 100. Composition of two superimposed waves

every value of the abscissa $t$, has as its ordinate the algebraic sum of the ordinates of the separate sinusoids at the same abscissa.

If the periods $T_i$ are commensurable, the resultant motion also is harmonic. For it to be a simple harmonic motion, however, special relations must prevail between the elements of the component motions. Only compositions of harmonic motions of equal period $T$ are of interest for our purpose.

If such component motions all have the same phase $\varphi$, then

$$s = (A_1 + A_2 + \cdots + A_n) \sin \left( \frac{2\pi}{T} t - \varphi \right)$$

The resultant is a simple harmonic motion, whose period and phase are equal to those of the components, and whose amplitude is equal to the algebraic sum of their amplitudes. If two such component motions have equal amplitude $A$, the resultant motion has an equal period, equal phase, and twice the amplitude:

$$s = 2A \sin \left( \frac{2\pi}{T} t - \varphi \right)$$

If the two component motions are of equal amplitude but in opposite phase, the resultant is complete rest:

$$s = 0$$

If the two component motions differ in amplitude and phase

$$s_1 = A_1 \sin (\omega t - \varphi_1) \qquad s_2 = A_2 \sin (\omega t - \varphi_2)$$

where $2\pi/T$ is replaced by $\omega$ for the sake of simplicity, the resultant motion is described by the equation

$$s = s_1 + s_2 = (A_1 \cos \varphi_1 + A_2 \cos \varphi_2) \sin \omega t - (A_1 \sin \varphi_1 + A_2 \sin \varphi_2) \cos \omega t$$

This may be put into the form

$$s = A \sin (\omega t - \varphi) = A \cos \varphi \sin \omega t - A \sin \varphi \cos \omega t$$

provided that

$$A \sin \varphi = A_1 \sin \varphi_1 + A_2 \sin \varphi_2$$
$$A \cos \varphi = A_1 \cos \varphi_1 + A_2 \cos \varphi_2$$

Square these two equations and add them:

$$A^2 = A_1{}^2 + A_2{}^2 + 2A_1A_2 \cos (\varphi_2 - \varphi_1)$$

Divide the same two equations, member by member:

$$\tan \varphi = \frac{A_1 \sin \varphi_1 + A_2 \sin \varphi_2}{A_1 \cos \varphi_1 + A_2 \cos \varphi_2}$$

The resultant is also a simple harmonic motion, whose period is equal to that of the component motions, and whose amplitude $A$ and phase $\varphi$ are defined by the last two equations above.

These equations show that the amplitude $A$ may be considered the resultant of the amplitudes $A_1$ and $A_2$, represented by two vectors that have lengths proportional to the respective magnitudes

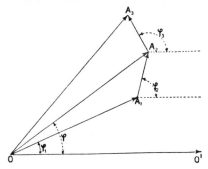

FIG. 101. Composition of wave motions of equal wave length

and that form between them an angle equal to the phase difference $(\varphi_1 - \varphi_2)$. If there are $n$ component motions, the simple harmonic motion resulting from them all is found by repeating $(n - 1)$ times the operation just described.

A very valuable graphic procedure directly utilizes the foregoing principle. Given two motions $s_1$ and $s_2$, defined as above, choose a pole $O$ and a polar axis $OO'$, as in Fig. 101. Starting from $O$, draw a segment $OA_1$, proportional to $A_1$ according to an arbitrary scale, and inclined to the polar axis at an angle $\varphi_1$. Starting from $A_1$, draw a segment $A_1A_2$, proportional to $A_2$ according to the same scale, and inclined to the axis $OO'$ at an angle $\varphi_2$. The segment $OA_2$ represents the amplitude of the resultant motion on that scale, and the angle $\varphi$ that $OA_2$ makes with the polar axis indicates the phase of the resultant motion.

If we have to compound a third motion

$$s_3 = A_3 \sin (\omega t - \varphi_3)$$

with the first two, we need only start from $A_2$ and draw the segment $A_2A_3$ proportional to $A_3$. The amplitude of the resultant motion is given by $OA_3$; and the phase, by the angle $A_3OO'$. If other motions are to be compounded, the same procedure is followed. It also demonstrates at once the various cases considered previously in this Section.

**299.** When the trajectories of the component motions are no longer on the same straight line, the general problem of composition becomes rather complicated. We shall consider here only some very simple cases, in which the component motions have an equal period, trajectories at right angles to each other, and a common starting point.

When two Cartesian coordinates $x$ and $y$ are parallel to the two trajectories, the equations of the component motions are

$$x = X \sin (\omega t - \varphi_x)$$
$$y = Y \sin (\omega t - \varphi_y)$$

If the two motions are in phase ($\varphi_x = \varphi_y = \varphi$), the displacement $s$ of the moving point $P$ from the origin $O$ at any instant is given by

$$s = \sqrt{x^2 + y^2} = \sqrt{X^2 + Y^2} \sin (\omega t - \varphi)$$

The resultant is itself a simple harmonic motion whose amplitude is $\sqrt{X^2 + Y^2}$. Its trajectory is inclined to the $x$-axis at an angle $\alpha$ defined by the relation $\tan \alpha = Y/X$; in other words, the trajectory is the diagonal of the rectangle formed by the *positive* directions of the component motions (Fig. 102).

If the component motions are in opposite phase ($\varphi_y = \varphi_x \pm \pi$), everything comes out as in the preceding case, except that $\tan \alpha$ changes sign. This means that the trajectory of the resultant motion is the diagonal of the rectangle formed by one positive and one negative direction of the component motions.

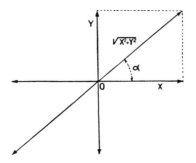

FIG. 102. Composition of wave motions in phase, when their trajectories are mutually perpendicular

If these are in quadrature $(\varphi_y = \varphi_x + \pi/2)$, we may write

$$x = X \sin (\omega t - \varphi_x)$$
$$y = Y \cos (\omega t - \varphi_x)$$

Squaring and adding, we have

$$\frac{x^2}{X^2} + \frac{y^2}{Y^2} = 1$$

This is the equation of an ellipse, whose center is at the origin and whose semiaxes are $X$ and $Y$ (Fig. 103). If $\omega t - \varphi_x = 0$, then

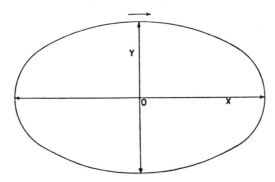

FIG. 103. Composition of two wave motions in quadrature, when their trajectories are mutually perpendicular

$x = 0$ and $y = +Y$; if $\omega t - \varphi_x = \pi/2$, then $x = +X$ and $y = 0$. Hence the ellipse is traversed clockwise, if the axes are disposed as in

Fig. 103. If the phase of one of the component motions were reversed, the direction of the vibration would be reversed too.

Modify the preceding case by adding the condition that the two amplitudes are equal $(X = Y = R)$; then

$$x = R \sin (\omega t - \varphi) \qquad y = R \cos (\omega t - \varphi)$$

Squaring and adding, we have

$$x^2 + y^2 = R^2$$

The trajectory is a circle; its radius is $R$, and it is traversed clockwise (Fig. 104a). Similarly, the composition of the motions de-

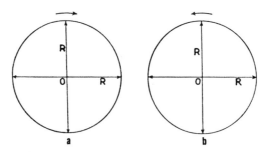

FIG. 104. Composition of mutually perpendicular wave motions in quadrature, when the two component motions are equal in amplitude

scribed by the following two equations

$$x = R \cos (\omega t - \varphi) \qquad y = R \sin (\omega t - \varphi)$$

leads to a circular trajectory, but it is traversed counterclockwise (Fig. 104b). In both cases the period of revolution is the same as that of the component motions, and the speed is uniform.

**300.** Most of the discussion in the foregoing Sections can be reversed, thereby permitting the immediate enunciation of the following two conclusions. In the first place, a simple harmonic motion described by the equation

$$s = A \sin (\omega t - \varphi)$$

and occurring along a trajectory inclined at an angle $\alpha$ to the

$x$-axis of a pair of rectangular coordinates $(x, y)$ intersecting in the starting point $O$ can always be decomposed into two simple harmonic motions occurring along the axes and described by the equations

$$x = (A \cos \alpha) \sin (\omega t - \varphi)$$
$$y = (A \sin \alpha) \sin (\omega t - \varphi)$$

Secondly, a *uniform* circular motion can always be decomposed into two simple harmonic motions in quadrature, their trajectories being mutually perpendicular, their amplitudes being equal to the radius of the circle, and their periods being equal to the period of the circular motion.

**301.** In connection with the decomposition of vibratory motions, *development in Fourier series* deserves to be mentioned, but only a simple enunciation of it will be made here. Of any curve $s(t)$, consider the part within the interval between the values $-T/2$ and $+T/2$ of the variable. Between these values let the curve satisfy certain mathematical restrictions, which are always fulfilled in the cases of interest to us. As usual, put $\omega = 2\pi/T$. Then Fourier showed that $s(t)$ between the values $s\left(-\dfrac{T}{2}\right)$ and $s\left(\dfrac{T}{2}\right)$ can be represented by

$$s(t) = \sum_0^\infty (a_l \sin l\omega t + b_l \cos l\omega t)$$

The series converges, and usually rather rapidly. The coefficients $a$ and $b$ may be calculated in practice by many methods, graphical included. There are also mechanical and electrical devices, called *harmonic analyzers*. Once the given curve has been introduced into them, they yield the principal coefficients of development in Fourier series.

In the special case when complex periodic motions are to be decomposed, the development must be carried out over a range of values corresponding to a period.

The function $s(t)$ is thus produced by the composition of many sinusoids, of which the one with the longest period is called the *fundamental harmonic*, while the others are termed "harmonics of higher order."

**302.** For a brief glance at the laws pertaining to the dynamics of harmonic motion we may begin with the simple equation

$$s = A \sin \frac{2\pi}{T} t$$

since the phase refers only to a displacement of the phenomenon in time, and that displacement has no dynamic effect. The velocity at any instant is given by

$$v = \frac{ds}{dt} = \frac{2\pi A}{T} \cos \frac{2\pi t}{T}$$

and the acceleration by

$$a = \frac{dv}{dt} = \frac{-4\pi^2 A}{T^2} \sin \frac{2\pi}{T} t = -\frac{4\pi^2}{T^2} s$$

If the moving point has a mass $m$, in harmonic motion it is subject to a force $f$, which varies in time according to the law

$$f = ma = \frac{-4\pi^2 m}{T^2} s$$

The force is, therefore, proportional to the displacement $s$ and is exerted in the opposite direction.

The energy of the moving point consists of two terms, one of which stands for the kinetic energy $E_k$, and the other for the potential energy $E_p$. The former is easily calculated by introducing into the formula $mv^2/2$ the expression found above for the velocity

$$E_k = \frac{2\pi^2}{T^2} mA^2 \cos^2 \frac{2\pi t}{T}$$

The potential energy $E_p$ in any point $P$ on the trajectory is equal to the work performed in transporting the moving point from $O$, the position of rest, to the point $P$. Since we are dealing with a continuously variable force, we divide the trajectory into innumerable elements $ds$, for each of which we take the product $-f\,ds$ (since $f$

and $s$ are always of opposite sign), and then we sum the elementary products by integration

$$E_p = -\int_o^s f \, ds = \frac{4\pi^2}{T^2} m \int_o^s s \, ds = \frac{2\pi^2}{T^2} ms^2 = \frac{2\pi^2}{T^2} mA^2 \sin^2 \frac{2\pi t}{T}$$

The total energy of the moving point is given by

$$E = E_k + E_p = \frac{2\pi^2}{T^2} mA^2$$

Therefore in motions of equal period it is proportional to the square of the amplitude.

# CHAPTER VIII

# *The Propagation of Waves*

◇◇◇◇◇◇◇◇◇◇◇◇

**303.** Let us imagine that a material point $P$, consisting of a minute particle, is executing a simple harmonic motion. If the particle is not alone, but is connected with other similar particles or, for the sake of simplicity, equal particles, its motion is accompanied by the phenomenon known as the *propagation of waves*. For if the various adjacent particles are tied together, once the point $P$ is set in motion, it has to drag the others more or less behind it. In this way its motion must be propagated in its vicinity.

Let us begin with the simplest case, that of particles linked in a line so as to form a string (Fig. 105a). To avoid certain complications that will be discussed later (§ 307), suppose $O$ is an accessible end, while the other end is very far away. Give $O$ a simple harmonic motion described by the equation

$$s = A \sin \frac{2\pi}{T} t$$

and let the direction of the motion be perpendicular to the string. Another particle $O_1$ immediately alongside of $O$ also moves, following the first particle, but with a certain lag, which may be very slight. For if there were no lag between $O$ and $O_1$, neither would there have to be any lag between $O_1$ and a subsequent particle $O_2$, and so on. Then when the point $O$ went down, the whole string would have to go down, while remaining straight and parallel to its initial position. But this never happens, because there are no absolutely rigid strings, at any rate if we take them long enough.

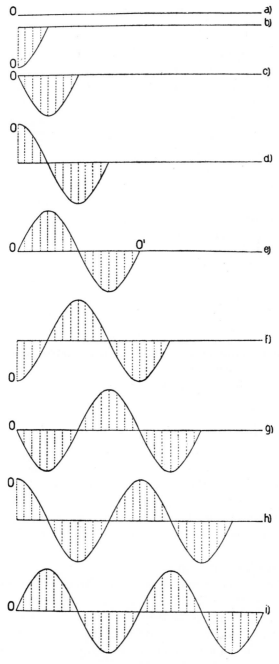

FIG. 105. Propagation of waves
311

Hence it follows that the particle $O_1$ moves with the same motion as $O$, but lags a little behind it; $O_2$ does the same with respect to $O_1$; and so on. In Fig. 105 consider various of these particles, each at a constant distance from its predecessor. Start from the instant $t = 0$, when $O$ begins to vibrate. After an interval $t = T/4$, the string looks like the curve in Fig. 105$b$. After $t = T/2$, the curve $c$ shows us the form of the string; $d$, after $t = 3T/4$; and $e$, after $t = T$. At this instant the point $O$ has returned to its starting position and, if its motion is continued, it is ready to begin its second cycle. At the same instant another particle $O'$ commences its motion in the same kinematic circumstances as $O$, and the motions of these two particles will always be identical and in phase. Within the interval lying between them, it should be noted, there is no other particle which fulfills these conditions. The interval $OO'$ is in general the *shortest* distance between two points vibrating in phase; it is called the *wave length* and is denoted by $\lambda$.

Meanwhile the motion is being propagated along the string. Its form after a time $t = 5T/4$ is indicated by Fig. 105$f$; after a time $t = 6T/4$, by $g$; when $t = 7T/4$, by $h$; when $t = 2T$, by $i$; and so on. It is clear that the motion is propagated, while the individual particles have only oscillated around their positions of equilibrium. This phenomenon is expressed by saying that *waves are propagated* or travel along the string.

If the latter is homogeneous, the lag with which the motion is transmitted from one particle to the next is constant, and therefore $V$, the *velocity of propagation* of the waves along the string, is constant. Consequently the motion of the waves is uniform. The velocity $V$ is entirely distinct from the velocity of the harmonic motion of the individual particles.

The phase is the only kinematic characteristic that distinguishes the motion of the point $O$ from the motion of the other points. Each of these accordingly may be considered as the source of the vibrations transmitted to the subsequent points.

When there is no friction or it is disregarded, all the particles of the string execute harmonic motions that are identical even in amplitude. If we compute the time for each particle from the instant when it is set in motion, the equation of the motion is the same for all of them as it is for the point $O$. But if in the equation for the motion of each particle we reckon the time as commencing from the instant when the point $O$ starts to move, then the equation must take the following form, previously given in § 296,

$$s = A \sin \frac{2\pi}{T}(t - \tau)$$

where $\tau$ stands for the lag after which the point referred to in the equation begins to vibrate as compared with $O$. In other words, $\tau$ is the time needed by the wave to traverse the distance $x$ from $O$ to that point.

As was remarked just above, the propagation of the waves is a uniform motion. Hence

$$\tau = \frac{x}{V}$$

and therefore

$$s = A \sin 2\pi \left(\frac{t}{T} - \frac{x}{TV}\right)$$

Given the definition of the wave length $\lambda$, we see at once that the waves are propagated over a distance equal to $\lambda$ during the time $T$ in which the source performs one complete oscillation. Consequently

$$\lambda = VT$$

and at the same time

$$s = A \sin 2\pi \left(\frac{t}{T} - \frac{x}{\lambda}\right)$$

In this equation we may consider both $t$ and $x$ as variables. If $t$ is given a fixed value, the function $s = f(x)$ gives us the form assumed by the string at the instant $t$ in the rectangular coordinates $s$ and $x$. Obviously this form is a sinusoid, as is shown by Fig. 105.

If the value of $x$ is fixed, the variable is $t$ and the equation above becomes the equation of the motion of that point on the string which is at distance $x$ from $O$. All the points at distance $\lambda$ or $k\lambda$ ($k = \pm0, 1, 2, \ldots$) complete motions that are identical even in phase, as is readily demonstrated.

**304.** In the preceding Section we considered a displacement $s$ of the various particles in a direction perpendicular to that of the string. When this happens, the vibration is said to be *transverse.* The two directions may also coincide, and then the vibration is called *longitudinal;* or they may form any angle between them, and then the vibration can be decomposed into a longitudinal and a transverse vibration. We shall not tarry long over this topic, however, because only transverse vibrations are of interest for our purpose.

In the preceding Section we considered each particle as executing a simple harmonic and therefore rectilinear motion. But each

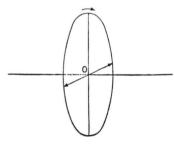

particle may also perform a uniform circular motion in a plane perpendicular to the string (Fig. 106). This case is quickly reduced, however, to that of two simultaneous simple harmonic motions in quadrature, whose amplitudes are equal and whose directions are perpendicular to each other and to the string (§ 299). If each particle moved in an elliptical trajectory

FIG. 106. Transverse wave circularly polarized

perpendicular to the string, the decomposition would lead to two simple harmonic motions of different amplitude (§ 299).

In general, whatever the periodic motion of any point on the string may be, it can be developed in Fourier series, and thus will be decomposed into a group of transverse simple harmonic motions, and possibly into another group of longitudinal motions. In practice, the vibrations with which we shall deal are complex in the general case. To describe some simple cases, certain expressions have been introduced.

When the points of the string execute a transverse simple harmonic motion, the resulting wave always lies in one plane and is said to be *linearly polarized* in the plane perpendicular to the plane in which the particles of the string move. When the motion of those particles is transverse and either circular or elliptical, the wave is said to be *circularly* or *elliptically polarized.*

So far as energy is concerned, for the propagation to occur, obviously energy must be supplied continuously. For if the particle $O$ is not kept moving by continuous external work, it surrenders its energy to the adjacent particles, and when it has exhausted its energy, it stops moving. The wave is then confined within a brief interval, called a *wave train*, which travels along the string. The length of this wave train is measured by the number of wave lengths corresponding to the number of oscillations completed by the first particle *O*. When this number is as big as you like, the waves are called *persistent*. On the other hand, if the number is small, the waves are termed *damped*. These last are of no interest to us.

**305.** The concepts set forth in the two preceding Sections may be extended, with appropriate modifications, to the propagation of waves on a surface.

If a particle *O* is set in motion and kept moving, it transmits its motion at the same time to a whole lot of other particles all around it. Each of these does as much. Hence in any direction lying on the surface (which we shall assume to be plane) and passing through *O*, what happens is like what we found happening along a string. Given symmetry around the point *O*, when the surface under consideration is *homogeneous*, the waves assume the aspect of numerous concentric circles with center at *O* and continuously increasing radius. A familiar example would be the rings generated by striking the surface of a liquid at rest.

The energy that is distributed along a circle later passes along a circle of bigger radius and therefore is distributed among a larger number of particles. Since the surface has been assumed to be homogeneous, the number of particles distributed along a circumference is proportional to its length, or to the radius $x$. Consequently the energy of each particle is proportional to $1/x$. The period of oscillation is identical for all the points, and the particles are all equal and hence of equal mass. Accordingly the motion's amplitude, the square of which is proportional to the energy (§ 302), must decrease as $x$ increases, according to the law

$$A^2 = \frac{C}{x}$$

where $C$ stands for the square of the amplitude of the motion of the points whose distance from the origin $O$ is $x = 1$. For $x = 0$, however, this law has no meaning.

A similar analysis may be made for the case in which the point $O$ transmits its motion in three dimensions. If the medium in which it is immersed is homogeneous and isotropic, concentric spherical waves are propagated around the origin.

This extension is a bit too simple, however, and to be understood better would require a detailed examination of the case being considered. Suppose a particle executes a linear simple harmonic motion. In the surrounding space establish three rectangular coordinates $x, y, z$, with the $z$-axis running in the same direction as the trajectory of the oscillating motion. Clearly, in every direction lying in the $xy$ plane transverse waves are propagated. But in the direction of the $z$-axis there are only longitudinal waves. In the intermediate directions there are complex motions, one component being transverse and the other longitudinal.

As was pointed out in § 304, only transverse waves are of interest to us. Hence we shall have to confine our discussion to waves propagated in very small solid angles.

So far as the distribution of energy in the waves is concerned, an argument similar to the one stated above shows that the energy of the particles distributed over a spherical surface passes to the particles of a subsequent spherical surface of bigger radius $x$. Since the density of the particles is constant, their number is proportional to the surface of the sphere and therefore to $x^2$. Hence the amplitude of the motion of the individual particles diminishes according to the relation

$$A^2 = \frac{C}{x^2}$$

where $C$ is a constant equal to the square of the amplitude of the motion of the particles whose distance from the origin is $x = 1$.

Hence is derived the law that in the propagation of spherical waves the amplitude of the motion decreases in inverse proportion to the distance from the source or, what amounts to the same thing, in the propagation of spherical waves the energy of the motion diminishes inversely as the square of the distance from the source.

**306.** In connection with the propagation of waves along a string it was pointed out in § 303 that, except for the phase, there is no kinematic difference between the motion of the source $O$ and the motion of any other particle $P$. Consequently a particle $P'$ subsequent to $P$ may be considered as being set in motion by a wave whose source is in $P$.

A similar argument may be repeated for the propagation of waves over a surface. Every particle, once it has been set in motion by the arrival of a wave, becomes in turn a source of waves, *almost* like the central source $O$, because the particle's motion is necessarily transmitted to the adjacent particles.

As an example of such propagation, attention was called almost three centuries ago to the way in which a fire spreads. The periphery

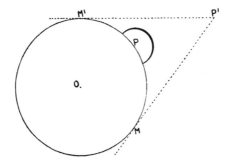

Fig. 107. Principle of secondary wavelets

of the burning area ejects sparks which, when hurled forward, kindle local fires, each on its own account, just as the first spark had done in starting the principal conflagration.

Yet the analogy between these new peripheral sources of waves and the central source is not as close as in the propagation of waves along a string. For not only does the phase change, but also the amplitude (§ 305). Moreover, these peripheral sources must absorb energy on the side facing the central source and emit it only on the opposite side, whereas the point $O$ receives energy from outside the surface and emits it in all directions on the surface.

Therefore all the points of a circular wave around a source $O$ have to be regarded as being in their turn sources of semicircular waves moving outward (Fig. 107). Choose a point $P'$ outside the circle and at a distance $x'$ from $O$. The motion of $P'$ may be treated as resulting from the composition of the motions reaching the point

$P'$ from all the points vibrating, each on its own account, along the arc $MM'$ of the wave under discussion.

If the whole arc $MM'$ is in operation, the effect on $P'$, as we already know, is expressed by

$$s' = A' \sin 2\pi \left( \frac{t}{T} - \frac{x'}{\lambda} \right)$$

where $A'$ is an amplitude suitably different from the amplitude $A$ of the point $O$, as was shown in § 305. But if the arc $MM'$ is not homogeneous, or if obstructions modify the progress of the waves, the effect on $P'$ must change and can no longer be represented by the equation above.

The waves emitted by the individual elements $P$ of the wave under consideration may be called *elementary* or *secondary wavelets*. The method just explained for determining the motion of the point $P'$ is contained in the following rule. The motion in a point $P'$ outside a wave is the integral of all the motions generated in the point individually by all the secondary wavelets emitted by the points of the arc $MM'$. This is the *principle of secondary wavelets*, which forms the basis of the theory of wave propagation.

The idea of secondary wavelets has an obscure origin. It was applied in the study of mechanical waves, especially acoustical. But in the field of visual waves it received at the hands of Christiaan Huygens a remarkable geometrical interpretation, as a result of which it is often called *Huygens' principle*.

He noted that the individual secondary wavelets must be extremely weak, and therefore cannot be felt at some distance (on account of the normal decrease in their amplitude). Hence a wave that can be detected even at very great distances must be viewed as the joint effect of countless secondary wavelets. For Huygens, who treated this subject exclusively in a geometrical manner, the decisive wave was the *envelope of the secondary wavelets*. Thus a wave would be perceptible only where an envelope was possible. For elsewhere, if the individual secondary wavelets remained independent even though mixed together, the total effect would be imperceptible.

From this mechanism Huygens derived very important consequences, which are still widely repeated nowadays because they are simple intuitions. The concept of phase, however, was not explicitly introduced into this discussion, which is therefore not entirely conclusive. In other words, no claim is made that the conclusion is rigorous. It is an intuition, confirmed later by more complete calculations.

The formulation of the *principle of the integration of the secondary wavelets*, as set forth above, is due to Augustin Fresnel at the beginning of the nineteenth century.

To come closer to the experimental findings, we must regard the secondary wavelets as semicircular rather than circular. We must

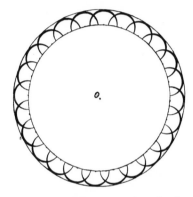

Fɪɢ. 108. Propagation of a circular wave, according to the principle of the envelope of secondary wavelets

Fɪɢ. 109. Propagation of a plane wave, according to the principle of the envelope of secondary wavelets

also recognize that in a semicircular wave the amplitude decreases from a central maximum to zero at the two extremities. Hence the maximum amplitude is found in the direction in which the joint wave is propagated. To make this fact clear, Fig. 107 shows the secondary wavelet thicker in the central portion of its semicircumference.

When a source emits a wave that at a given instant reaches a line all of whose points vibrate in phase, the line is called a *wave front*.

The propagation of a circular wave as an envelope of secondary wavelets is schematically illustrated in Fig. 108, and Fig. 109 does the same for a plane wave.

The group of arguments and ideas presented above may be extended to waves in space. Figs. 107, 108, and 109 are then interpreted as sections of spherical or plane waves. The secondary wave-lets are, therefore, semispherical, and their amplitude decreases toward the circular rim, where it vanishes in that plane tangent to the wave in which the elementary source lies. The wave front is now a surface whose points are reached by a wave at the same instant, and consequently vibrate in phase. The principle of the secondary wavelets may be expressed in the following general form. So far as concerns the effect exerted on an external point, sources of waves inside a closed surface may be replaced by sources on the surface.

# CHAPTER IX

# *Reflection, Refraction, and Diffraction of Waves*

◇◇◇◇◇◇◇◇◇◇◇◇

**307.** In the preceding Chapter we examined the propagation of waves along homogeneous and endless strings, over homogeneous and unlimited surfaces, in homogeneous and unlimited media. Now let us introduce a discontinuity in the progress of the waves. For example, let us interrupt the string along which they are propagated, by leaving one end free or tying it to a support; let us put an obstruction in the path of the waves propagated over a surface or in a three-dimensional medium.

Begin with the case of an interrupted string, tied to a support. Its last particle, being fixed, cannot move in response to the action of the preceding particles. For this immobility to occur, however, the support must react at every instant with a force equal and opposite to that action. The support therefore exerts a force varying sinusoidally. Hence ensues a harmonic motion, which is propagated along the string in the direction opposite to that from which the original motion arrived at the support.

The arriving wave is called the *incident wave;* the one that turns back is known as the *reflected wave.* The phenomenon is termed *reflection of waves* by an obstruction. If this absorbed no energy, the reflected wave would have the same amplitude as the incident wave. Moreover, the periods of both waves are obviously identical.

If the string is suspended from above and has its lower end free so that it dangles, reflection still takes place. The last particle receives energy from its immediate neighbor, and being unable to pass this energy on to any subsequent particle, restores it to the neighbor, thereby generating a new wave in the direction opposite to that of the arriving wave. The new wave is a reflected wave, which is the regular continuation of the incident wave.

On the other hand, if the end of the string is tied to a support, the reflected wave is in opposite phase to that of the incident wave. Everything happens as if half a wave length were lost (or gained) in the reflection at the obstruction.

**308.** Let us turn to waves on a surface. Consider a rectilinear wave front $ABCD$ at the instant in which it touches a rectilinear

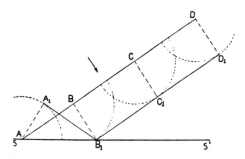

FIG. 110. Reflection of a plane wave by a plane mirror after a time $t_1$

obstruction $SS'$ at $A$ (Fig. 110). Apply the principle of the envelope of secondary wavelets in the manner explained by Huygens.

The point $A$ is on the obstruction. Therefore its secondary wavelets cannot push forward but must turn back, for energy reasons similar to those set forth in § 307 with regard to strings. Hence after a time $t_1$, $A$ has propagated circular waves of radius $AA_1 = Vt_1$, $V$ being the velocity of propagation of waves on the surface under consideration. At the same time, points $B$, $C$, $D$ . . . have emitted secondary wavelets that have traversed a distance

$$BB_1 = CC_1 = DD_1 = \cdots = Vt_1 = AA_1$$

toward the obstruction. To locate the wave front after this interval of time $t_1$, we must trace the envelope of the secondary wavelets that have been generated.

The point $B_1$ is on the obstruction. The envelope from $B_1$ to $D_1$ is undoubtedly given by the straight line $B_1C_1D_1$, as we saw in § 306 and Fig. 109. The remaining portion must be the envelope of the secondary wavelets that have already been affected by the obstruction. Since the wavelet leaving from $A$ also forms part of the envelope, the wave front is obtained by drawing the tangent from $B_1$ to the wave $A_1$. In this way the remaining portion $B_1A_1$ is found. Consequently after the time $t_1$ the wave front is $A_1B_1C_1D_1$. The portion $A_1B_1$ is reflected, while the rest goes ahead to meet the obstruction.

After a time $t_2 = 2t_1$, reckoned as starting from the instant in which the wave front was at the position $ABCD$ in Fig. 110, we get

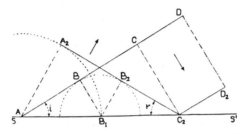

Fig. 111. Reflection of a plane wave by a plane mirror after a time $t_2$

a configuration like that shown in Fig. 111. The wavelets leaving from the segment $CD$ have as their envelope the front $C_2D_2$. Those leaving from the segment $ABC$, having been affected by the presence of the obstruction $SS'$, give rise to the envelope $A_2B_2C_2$, and the following equivalents must be true:

$$AA_2 = BB_1 + B_1B_2 = CC_2 = DD_2 = Vt_2 = 2Vt_1$$

Then the wave front becomes $A_2B_2C_2D_2$. The portion $A_2C_2$ is reflected, while $C_2D_2$ proceeds toward the obstruction. When the wavelet emitted by $D$ reaches the obstruction, the entire wave front is reflected and advances as a whole in the new direction perpendicular to the front.

The angle $i$, which the wave front $ABCD$ makes with the obstruction $SS'$, is called the *angle of incidence*. The angle $r$, formed by the reflected wave front and the obstruction, is termed the *angle of reflection*. The two right triangles $AA_2C_2$ and $ACC_2$, with a common hypotenuse $(AC_2)$ and one side equal by construction $(AA_2 = CC_2)$, have their corresponding angles equal, so that

$$i = r$$

This is the law of reflection of waves on a surface. The law implies that $AA_2$ and $CC_2$ are equal. But these two segments measure the progress of the reflected and incident waves, respectively, in the same interval of time. Hence the law of reflection asserts that the incident and reflected waves are propagated with equal velocity. This equality is a necessary consequence of the fact that both waves are propagated on the same sur-

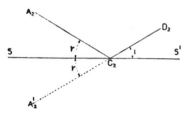

FIG. 112. Law of reflection of waves

face, for the velocity of propagation is a characteristic of the surface.

If the wave front is a curve, we repeat the foregoing analysis for each element of the wave. We soon reach the conclusion that at a given instant the wave front reflected by a rectilinear obstruction is the line symmetrical, with respect to the obstruction, to the line that would have been the wave front at the same instant, had the obstruction not been present. The same statement may of course be made about a rectilinear wave (Fig. 112).

**309.** The identical construction may be repeated for plane waves, or waves of any form, in a three-dimensional medium containing a plane surface of discontinuity. The results are similar. Not only is the law $i = r$ found to be valid, but considerations of symmetry lead to the conclusion that the two angles must lie in parallel planes.

Yet this construction is not rigorous. We need only notice that no mention has been made of what happens at the edges of the enveloping zone, whereas the phenomena occurring there are far from negligible. But we need not discuss them now.

On the other hand, we should pay attention to a concept that constitutes one of the basic rules of the wave mechanism. The advance of waves is constant; in other words, whenever the source completes one vibration, the wave front must go forward one wave length in all its elements. We shall see a remarkable application of this principle in § 316.

The surface of discontinuity by which the waves are reflected is called a *mirror;* if it is plane, it is termed a *plane mirror.*

**310.** In examining the behavior of a wave incident upon a *spherical mirror*, the following theorem is useful. Set up two rectangular coordinates $x$, $y$ in a plane (Fig. 113). A circle tangent to

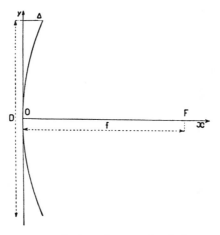

FIG. 113. Sagitta of an arc of a circle

the origin $O$, with radius $f$ and center $F$ on the $x$-axis, is described by the equation

$$(x - f)^2 + y^2 = f^2$$

Take a small arc of this circle, the coordinates of the ends of the arc being

$$x = \Delta \qquad y = \pm \frac{D}{2}$$

By substituting in the previous equation we get

$$f = \frac{4\Delta^2 + D^2}{8\Delta}$$

Provided that the arc is small enough, $4\Delta^2$ is negligible in comparison with $D^2$. Hence we have the following approximations:

$$f = \frac{D^2}{8\Delta} \qquad \Delta = \frac{D^2}{8f}$$

which we shall employ extensively.

**311.** Now consider a concave spherical mirror $MOM$, with radius of curvature $r$, diameter $D$, and vertex $O$ (Fig. 114). Let this

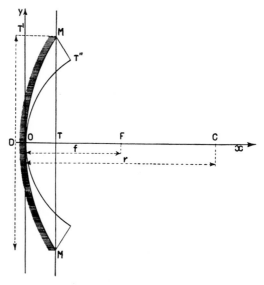

Fig. 114. Reflection of a plane wave by a spherical mirror

mirror be struck by a plane wave coming from the point at infinity on the axis $OC$, which we shall take to be the $x$-axis, with $C$ as the center of curvature. Starting from the moment when the mirror is struck by the incident wave, every element of the mirror becomes a center of secondary wavelets. In particular, the points $M$ are the first to emit wavelets, and $O$ is the last. The time by which $O$ lags behind $M$ in beginning to emit wavelets is exactly equal to the time taken by the incident wave to traverse the distance $TO = MT'$. Hence when $O$ commences to emit wavelets, those emitted by $M$ have already traversed a distance $MT'' = MT'$.

The envelope of the reflected secondary wavelets is obviously not spherical. But it may be said to have a form possessing symmetry of revolution around the axis $OC$. As a first approximation, however, we may consider it spherical, and then we can determine its radius of curvature $f$. The approximation consists of using the equation given for $f$ in § 310 and putting

$$MT' + MT'' = 2MT' = \Delta$$

By applying the last equation of § 310 to the section of the mirror, we get

$$MT' = \frac{D^2}{8r}$$

while for the reflected wave we have

$$f = \frac{D^2}{8\Delta} = \frac{D^2}{8 \times 2MT'} = \frac{r}{2}$$

Therefore the reflected wave has its center at the midpoint $F$ of the segment $OC$. This point is called the *focus* of the spherical mirror.

Now suppose that $S$, the source of waves, is still on the axis, but at a distance $x > r$ from the vertex of the mirror $MM$ (Fig. 115). When $M$ starts to vibrate, the incident wave is still at a distance $TO$ from $O$. But

$$TO = MT' - T'T'''$$

$T'''O$ being an arc of a circle whose center is at the source $S$ and whose radius is $x$. Using the equations of § 310 again, we have

$$MT' = \frac{D^2}{8r} \qquad T'T''' = \frac{D^2}{8x}$$

When $O$ begins to vibrate, the wavelet reflected by $M$ has already traversed a distance $MT'' = TO$. The reflected wave, which forms the envelope of the secondary wavelets, is generally not spherical. Yet it possesses symmetry of revolution around the axis. We may consider it as approximately spherical, with radius of curvature $x'$. The $\Delta$ corresponding to $x'$ is given by

$$\Delta = \frac{D^2}{8x'}$$

Using our previous approximation, we may write

$$\Delta = MT' + MT'' = 2MT' - T'T'''$$

and then, by substituting in this expression the values found above, we have

$$\frac{1}{x} + \frac{1}{x'} = \frac{2}{r} \qquad \text{or} \qquad \frac{1}{x} + \frac{1}{x'} = \frac{1}{f}$$

This equation lends itself to a very simple geometrical interpretation. The sum of the curvatures of the incident and reflected waves

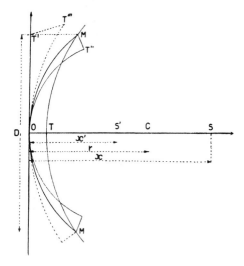

Fig. 115. Reflection of a spherical wave by a spherical mirror

is constant. If one curvature is zero, the other has the value $1/f$ and is twice the curvature of the mirror.

This equation is valid in general, whatever the value of $x$ may be. In order to understand its operation better, however, let us make a particular examination of the cases in which $x \leqq r$.

When $x = r$, the source is at $C$, the mirror's center of curvature. Then the waves strike all points of the reflecting surface at the same instant. Therefore the reflected waves are convergent spherical waves with their center at $C$. In other words, the equation is applicable because

$$x = x' = r$$

and therefore

$$\frac{1}{x} + \frac{1}{x'} = \frac{2}{r}$$

When $r > x > f$, the mirror is struck by the wave at the point $O$ first, and at the edge $M$ last. This case too is illustrated in Fig. 115, if the point $S'$ is considered as the source. Then $S$ becomes the center of the reflected waves. This fact by itself is enough to show that the computations are identical with those just above, and that therefore the final outcome is the same equation.

When $x = f = \frac{r}{2}$, we may return to Fig. 114 and consider the point $F$ as the source. Subject to the usual approximation, the wave reflected by the mirror is plane. Hence this case is covered by the equation under discussion.

When $x < f$, the reflected wave is convex; in other words, the center of curvature lies on the negative side of the $x$-axis. The equation under discussion still holds good, provided that the radius $x'$ of the reflected wave is given a negative sign.

**312.** The axis of a spherical mirror has little significance, since any straight line passing through $C$, the center of curvature, has a right to this name. It is nevertheless a common practice to consider the principal axis to be the straight line joining the center $C$ to the vertex $O$, the center of the mirror's edge.

In this sense we may speak of a point source of waves being situated outside the axis of a concave mirror. Let $S$ be such a source at a distance $y$ from the axis (Fig. 116). To find the center $S'$ of the reflected wave, extend the straight line $SC$ until it meets the mirror in $O'$. Clearly $S'$ must lie on this straight line. For if we diaphragm the mirror so as to reduce it to a small disk with its center at $O'$, then the straight line $SCO'$ is indeed the axis and $S'$, the center of the reflected waves, must lie on $SCO'$, as was shown in § 311. $S'$ of course keeps its position when the diaphragm is taken away.

Let $y'$ be the distance of $S'$ from the axis. Subject to the usual approximations, the equation

$$-\frac{y'}{y} = \frac{x'}{x}$$

holds true. To see this, we need only imagine the mirror so diaphragmed as to be reduced to a tiny disk around $O$. The small
waves projected by $S$ upon $O$ at angle $\alpha$ are reflected in the direction

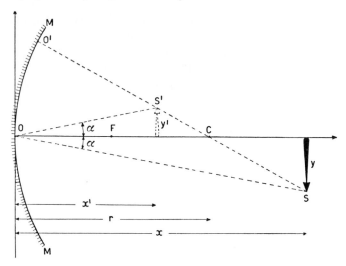

FIG. 116. Reflection of waves emitted by a point source not on the principal axis of a spherical mirror

$OS'$, which is inclined to the axis also at angle $\alpha$, in accordance with
the law of reflection. But

$$-\frac{y}{x} = \tan \alpha \qquad \frac{y'}{x'} = \tan \alpha$$

and the above equation follows immediately.

**313.** Let us turn back to consider waves propagated along a
string that is neither the homogeneous and endless string of Chapter
VIII nor the homogeneous and limited string of § 307. Our present
string consists of two parts, differing in structure and joined at a
point $B$ (Fig. 117). Let the second part be endless in the other
direction.

A wave leaves the source $S$ and reaches the point $B$. There the
particle that constitutes the last element of the part $SB$ transmits its
motion to the first element of the part $B\infty$. The velocity of propa-

gation depends on the delay in the transmission of the motion from one particle to the next. This delay in turn depends on the elastic connection between the particles or on the nature of the substance of the string. Hence the wave in the part $B\infty$ generally has a velocity $V'$ different from the velocity $V$ that it had in the part $SB$. But the period of vibration $T$ cannot change. For the particle at $B$ that serves as the source for the part $B\infty$ oscillates with the period of the last particle of the part $SB$, namely, the period of the source $S$. Therefore the period of oscillation cannot change at $B$, and consequently it cannot change at any other point on the string.

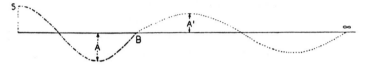

FIG. 117. Transmission of a wave along a nonhomogeneous string

This conclusion leads to the inference that the wave length $\lambda'$ in the part $B\infty$ must be different from the wave length $\lambda$ in the part $SB$. For, as we saw in § 303,

$$\lambda = VT \qquad \lambda' = V'T$$

and therefore

$$\frac{\lambda'}{\lambda} = \frac{V'}{V}$$

**314.** The question of the amplitude is more complex, because it is bound up with energy factors.

Assume that the particle at $B$ in the part $SB$ has a greater mass than the first particle of the part $B\infty$. Since both these particles are connected, their motion cannot differ in amplitude. Hence all the energy cannot pass from the first particle to the second, and therefore some of it must turn back (§ 307).

Now suppose that the ratio of the masses is reversed. The first particle of the part $B\infty$ cannot be displaced through an amplitude equal to the amplitude that the preceding particle would attain if it were unconstrained (in other words, the amplitude of the source $S$) because not enough energy is available. Therefore the amplitude of the motion in the part $B\infty$ must be smaller than the amplitude of the source $S$. Yet the last particle of the part $SB$ cannot differ in amplitude from the first particle of the part $B\infty$, because they are connected. Hence it follows that the last particle is, as it were, constrained. This constraint involves a sinusoidal reaction, which conveys a reflected wave from $B$ toward $S$ (§ 307).

We conclude then that at $B$ not only is there a transmission of waves from the part $SB$ to the part $B\infty$, but there must also be a reflection of waves from $B$ toward $S$.

The mechanism explained above does not pretend to represent the phenomenon completely. For real strings are not made up of individual particles lying in a straight line and connected by elastic forces, as we have supposed, but they have a much more complicated structure. Hence we shall limit our conclusions to the statements that this reflected wave must exist, and that it must be of greater magnitude, the bigger the difference between the characteristics of the two pieces of string.

**315.** If waves propagated over a surface encounter a zone in which the material structure of the surface is different from that of the preceding zone, the velocity of propagation must change (§ 313). Since the period of vibration remains unaltered, the wave length must vary in proportion to the velocity of propagation. The reasons given in § 314 to explain why there must be a reflected wave along strings apply also to the case of waves on a surface.

In addition, another very important phenomenon now manifests itself. In crossing the boundary line between the two zones of different nature, the wave generally deviates from the direction in which it originally advanced. This phenomenon, which is called the *refraction of waves*, may be explained intuitively, in Huygens' manner, by applying the principle of the *envelope* of secondary wavelets, as was done for the reflection of waves on a surface (§ 308).

Consider once more a rectilinear wave front $AB$ (Fig. 118). It is advancing toward $SS'$, the boundary line between two areas of different nature. Assume $SS'$ to be a straight line. Let us examine the behavior of the secondary wavelets, starting from the instant in which $A$, an extremity of the front, touches the line $SS'$.

The secondary wavelet whose center is $B$ advances in an area in which the velocity of propagation is $V$. After a time $t$, such that

$Vt = BB_1$, this wavelet reaches the point $B_1$ on $SS'$. During the same period of time $t$, the wavelet leaving from $A$ advances in the area of different nature. Consequently it has a different velocity $V'$, and traverses a distance $AA_1 = V't$, which is different from $BB_1$.

The new wave front must be the envelope of all the secondary wavelets formed in the space between $A$ and $B_1$. Hence the new front must be tangent to the two circumferences whose centers are $A$ and $B$, and whose radii are $AA_1$ and $BB_1$, respectively. Therefore the new wave front is the segment $A_1B_1$.

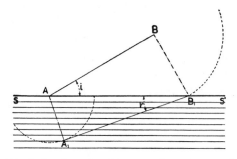

FIG. 118. Refraction of a plane wave by a plane surface

Its inclination $r$ to the line $SS'$ is connected with the inclination $i$ of the incident wave front $AB$ to the same line by the relation

$$\frac{\sin i}{\sin r} = \frac{V}{V'} = n$$

as is deduced from the triangles $ABB_1$ and $AA_1B_1$, in which

$$AB_1 = \frac{Vt}{\sin i} = \frac{V't}{\sin r}$$

Hence the ratio of the sines of the angles $i$ and $r$ is a constant, depending only on the velocities with which waves are propagated on the two areas. Since these velocities depend on the nature of the substances bounded by the surfaces, $n$ is a specific constant. The name *angle of incidence* is given to $i$; *angle of refraction*, to $r$; *refraction of waves*, to the phenomenon, and *index of refraction*, to $n$. The expression

$$\frac{\sin i}{\sin r} = n$$

is the law of refraction. Many people call it Descartes' law, because he was the first to publish it, in 1637; but its content, in somewhat different form, had already been discovered by Snel.

When either the wave front or the boundary line between the two different areas is not rectilinear, the law is applied to the constituent elements. There are, however, certain reservations to be made, which we need not discuss in detail.

**316.** The arguments and conclusions of §§ 313–315 are readily extended to the case of waves in space, when a wave front strikes a surface of separation $SS'$ between two media in which the wave is propagated with different velocities. Fig. 118 may be viewed as a section of such media, the section being made by a plane perpendicular to the wave and to the surface.

The usual reasons of symmetry require the angles $i$ and $r$ to be in parallel planes. Therefore waves in space are likewise subject to the law of refraction in the form

$$\frac{\sin i}{\sin r} = n$$

For the sake of greater precision, $n$ is called the *refractive index of the second medium with respect to the first medium*. In practice it is also customary to talk about the *refractive index of a substance*. This is a conventional way of speaking when the first medium is a vacuum. The refractive index that is found in this way is described as "absolute."

The above law may be applied to elements when the wave front or the surface of separation is curved.

In Fig. 118 the distance $AA_1$ differs geometrically from $BB_1$, since

$$\frac{AA_1}{BB_1} = \frac{V't}{Vt}$$

But, as we saw in § 313,

$$\frac{\lambda'}{\lambda} = \frac{V'}{V}$$

Hence

$$\frac{AA_1}{\lambda'} = \frac{BB_1}{\lambda}$$

Thus if we measure by wave lengths, we discover that both distances $AA_1$ and $BB_1$ contain exactly the same number of wave lengths. Moreover, this equality is a necessary consequence of the way in which waves are propagated, and of the definition of a wave front. The deviation undergone by a wave in crossing $SS'$, the surface of separation, may be otherwise expressed. We may say that all the elements of the front must advance by one $\lambda$ whenever the source completes one oscillation, that is, in the time $T$. Hence the front advances farther where the waves are longer. It can remain parallel to itself (or undeviated) only when the waves keep the same length along the entire front, or undergo equal variations in every element of the front.

This last case occurs when at the same instant all the elements of the wave front cross a surface of separation that has the same form as the front, and the incidence is perpendicular at every point.

To conclude this discussion, we should refer to the relation connecting the wave length, velocity of propagation, and index of refraction. Consider two media, in the first of which the wave length is $\lambda$ and the velocity of propagation is $V$, while in the second they are $\lambda'$ and $V'$. If $n$ is the refractive index of the second medium with respect to the first, we have

$$\frac{\lambda}{\lambda'} = \frac{V}{V'} = n$$

and therefore

$$\lambda = n\lambda' \qquad \text{or} \qquad \lambda' = \frac{\lambda}{n}$$

**317.** Now consider a medium separated from a surrounding medium by two plane surfaces inclined to each other at an angle $\alpha$ (Fig. 119). If a third surface (which is generally plane) further limits the two media so as to form three parallel edges, a *prism* results. Let the substance of which the prism is made have a refractive index $n$ with respect to the medium outside. Let $AB$ denote the incident wave front, and $A'B'$ the emergent wave front. Let $\lambda$ be the wave length in the outside medium, and $\lambda'$ in the prism.

For the reasons indicated in the latter part of § 316, the broken line $AVA'$ must contain the same number of waves as the line $BB'$, since $AB$ and $A'B'$ are two successive fronts of the same wave. But $AVA'$ is entirely in the outside medium, and $BB'$ is entirely in the prism. Hence

$$\frac{AV + VA'}{\lambda} = \frac{BB'}{\lambda'}$$

If $n > 1$, and therefore $\lambda > \lambda'$, $AV + VA'$ must also be greater than $BB'$. In other words, the wave must be deviated; and the

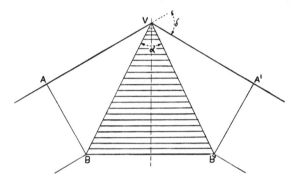

FIG. 119. Deviation of a plane wave by a prism

deviation must be away from the *refracting edge* $V$ or toward the *base*, the name given to $BB'$.

The angle **δ,** which the perpendicular to the incident wave forms with the perpendicular to the emergent wave, is produced by the refraction through the prism and is called the *angle of deviation* of the wave.

If $n$ were less than 1, the deviation would clearly take place toward the refracting edge rather than toward the base.

**318.** We saw in § 313 that the velocities with which waves are propagated in various media are functions of the nature of these media; and when we were introduced in § 315 to the index of refraction as the ratio between two velocities of propagation, we noticed that it is a *specific constant*. But experience shows that it is also a function of $\lambda$; in other words, the velocity of propagation of a wave motion in a medium generally varies as $\lambda$ varies. This phenomenon is called *dispersion* of the waves, a name whose origin is to be sought in the effect that the phenomenon has on the behavior of a prism.

We learned in § 317 that a wave that passes through a prism undergoes a deviation δ, but now we should inspect a little more carefully the conditions in which this phenomenon occurs.

Assume that the incident wave is pure, i.e., that it consists of a simple harmonic motion. Then it has a well-defined λ in the outside medium, and also a well-defined λ' inside the prism, the exact relation between the two being

$$\lambda' = \frac{\lambda}{n}$$

Hence there is a well-defined emergent wave, whose direction is inclined to the direction of the incident wave in accordance with the equation given in § 317:

$$\frac{AV + VA'}{\lambda} = \frac{BB'}{\lambda'}$$

or

$$AV + VA' = nBB'$$

Now assume that the incident wave consists of two slightly different waves, whose wave lengths are λ and λ + dλ. Each of these

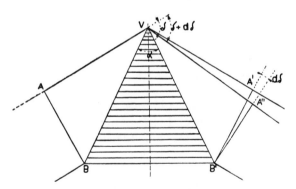

Fig. 120. Dispersion of waves by a prism

waves behaves differently inside the prism, because the refractive index of the substance of which the prism is made is n for the first wave and n + dn for the second. When the two waves emerge, they travel in different directions, since each has been deviated differently by the prism (Fig. 120). More precisely, if B'A' denotes the front of one emerging wave, and $B'A''$ the wave front of the other, we may write an equation like the last one above:

$$AV + VA'' = (n + dn)BB'$$

Subtracting these two equations, member by member, we get

$$VA'' - VA' = dnBB'$$

Let $D$ denote the average of the sections of the waves

$$D = \frac{B'A' + B'A''}{2}$$

Let $b$ stand for $BB'$, the base of the prism, and $d\delta$ for the small angle formed by the fronts of the two emerging waves. Then we may write

$$d\delta = \frac{b}{D} dn$$

since we are dealing with a very small angle. This equation is some-times used in the form

$$\frac{d\delta}{d\lambda} = \frac{b}{D} \frac{dn}{d\lambda}$$

The first term is called the *angular dispersion of the prism*, while $dn/d\lambda$ indicates the *dispersion of the substance* of which the prism is made.

**319.** Now let the two media be separated by a spherical surface with radius $r$ and center $C$. Let $O$ be the vertex, $OC$ the axis, and $D$ the diameter, as shown in Fig. 121. Let $S$, the source of waves, be in the first medium on the axis $OC$ and at a distance $x$ from $O$.

When the wave reaches $O$, that part of it that is at a distance $D/2$ from the axis lags behind by a segment $\delta_1$, which is given by the last equation of § 310 as

$$\delta_1 = \frac{D^2}{8x}$$

At the same distance $D/2$ from the axis, the spherical surface and the plane tangent to it at the vertex $O$ have between them a segment $\delta$, given by

$$\delta = \frac{D^2}{8r}$$

While the peripheral part of the wave is crossing the stretch $\delta_1 + \delta$ in the first medium, the central part of the wave advances in the second medium from $O$ toward $C$ over a space $\delta_2$. This stands in the

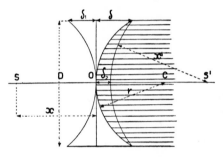

Fig. 121. Refraction of a spherical wave by a spherical surface

same ratio to $\delta_1 + \delta$ as the velocity $V'$ in the second medium to the velocity $V$ in the first medium

$$\frac{\delta_2}{\delta_1 + \delta} = \frac{V'}{V} = \frac{1}{n}$$

Thus in the second medium a wave is propagated whose central part no longer has the same lead over the peripheral part as it would have if its velocity had not changed. Whatever may be the form of the wave thus obtained (in any case it must have symmetry of revolution around the axis) we may consider it spherical, with a radius $x'$ given by

$$x' = \frac{D^2}{8(\delta - \delta_2)}$$

since the edge of the wave leads the paraxial part by a length of exactly $\delta - \delta_2$. From the preceding equations we obtain at once

$$\delta - \delta_2 = \delta - \frac{\delta_1 + \delta}{n} = \frac{(n-1)\delta - \delta_1}{n}$$

and therefore

$$\frac{1}{x} + \frac{n}{x'} = \frac{n-1}{r}$$

**320.** Now let the second medium be enclosed between two spherical surfaces, whose radii are $r_1$ and $r_2$, respectively. Let these two surfaces be so close to each other that $d$, the thickness of the medium, is negligible in comparison with the other magnitudes in question (Fig. 122). Let $D$ be the diameter of the second medium, and $n$ its

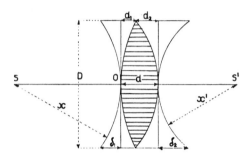

FIG. 122. Refraction of a spherical wave by a converging lens

refractive index with respect to the first medium. This arrangement constitutes a *thin lens.*

The last equation of § 310 gives us

$$d_1 = \frac{D^2}{8r_1} \qquad d_2 = \frac{D^2}{8r_2}$$

and since

$$d = d_1 + d_2$$

we get

$$d = \frac{D^2}{8}\left(\frac{1}{r_1} + \frac{1}{r_2}\right)$$

The system's axis of revolution is a straight line passing through the centers of curvature of the two spherical surfaces under consideration. Let $S$, the source of waves, be on this axis at a distance $x$ from the system. When the wave touches the second medium at $O$ on the axis, at the edge it is still a distance $\delta_1$ away from the plane that is tangent at $O$. Then

$$\delta_1 = \frac{D^2}{8x}$$

Although the emerging wave front is generally not spherical, we may consider it such, as a first approximation, and call its radius $x'$.

At the instant when it emerges from the second spherical surface at the axis, at the edge it is already a distance $\delta_2$ away from the plane that is tangent to this surface at its central point. Then

$$\delta_2 = \frac{D^2}{8x'}$$

According to the usual reasoning, in the time taken by the wave to cover the distance $d$ in the second medium, it must travel

$$\delta_1 + d_1 + d_2 + \delta_2$$

in the first medium. Therefore, by the results obtained in the preceding Sections,

$$\delta_1 + d_1 + d_2 + \delta_2 = nd$$

or

$$\delta_1 + \delta_2 = (n - 1)d$$

Substituting in this expression the values given by the preceding equations, we have at once

$$\frac{1}{x} + \frac{1}{x'} = (n - 1)\left(\frac{1}{r_1} + \frac{1}{r_2}\right)$$

In particular, if $x = \infty$, the common practice is to replace $x'$ by $f$. Then

$$\frac{1}{f} = (n - 1)\left(\frac{1}{r_1} + \frac{1}{r_2}\right)$$

and therefore

$$\frac{1}{x} + \frac{1}{x'} = \frac{1}{f}$$

It is not by accident that this last equation comes out identical with the equation obtained in § 311 for the reflection of waves by spherical surfaces. The reasoning in both cases is substantially the same.

Because equations of this type are used very frequently, it has been found convenient to define the *vergence* of a point with respect to some standard of reference as the reciprocal of the distance $x$ between them. The vergence is usually denoted by the Greek letter that corresponds to the English letter indicating the distance. Thus

$$\xi = \frac{1}{x} \qquad \xi' = \frac{1}{x'} \qquad \varphi = \frac{1}{f}$$

Then the above equation takes on the form

$$\xi + \xi' = \varphi$$

When the meter is employed as the unit of length, the vergence is expressed in *diopters*.

**321.** Now move the point $S$ to a distance $y$ from the axis of the thin lens (Fig. 123). The combination consisting of the source $S'$,

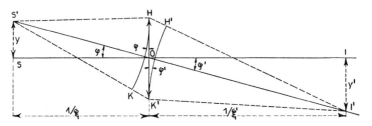

FIG. 123. Refraction of a spherical wave by a converging lens when the source is not on the axis

the lens $HOK'$, and the waves has lost the symmetry of revolution that prevailed under the conditions of § 320. This complication may be ignored, however, in our effort at a first approximation. Moreover, when the displacement $y$ is small, the variations in the form of the waves emerging from the lens are inappreciable, so that these waves may be considered to be still spherical.

Their center $I'$ can no longer be on the axis of the lens. But it must still be in the plane defined by the source and the axis, and that is a plane of symmetry. In Fig. 123 $y'$ is the distance from the center of the emergent waves to the axis $SI$. A mere line stands for the lens $HOK'$, since it is a thin lens whose thickness is negligible in comparison with the other magnitudes in question. Of course a lens with such a section cannot actually exist. For, as a result of the mechanism explained in § 320, the change in the curvature of the wave is due to the difference in thickness between the center of the lens and its edge. If anybody supposed that, just because the thickness can be considered negligible, therefore he should with greater reason ignore the differences in the thickness, he would eliminate the principal effect.

Under the conditions illustrated in Fig. 123, draw a circumference with center $S'$ and radius $S'H$, and another circumference with center $I'$ and radius $I'K'$. Clearly $HK$ represents the wave front incident upon the lens at the moment when the wave touches the lens at $H$, while $H'K'$ represents the emergent wave front at the instant when it leaves $K'$. Since these are two wave fronts, there must always be the same number of $\lambda$'s between $H$ and $H'$ as between $K$ and $K'$, because these two distances are always traversed in the same time; that is, $HH' = KK' = k\lambda$, if the lens is immersed in a homogeneous medium.

It follows at once that $y$ and $y'$ must be on opposite sides of the axis, a fact that can be indicated by giving them opposite signs. Denote the angle $S'OS$ by $\varphi$, and the angle $I'OI$ by $\varphi'$. Since we are dealing with small angles, we have

$$\varphi = \frac{y}{x} = y\xi \qquad \varphi' = \frac{y'}{x'} = y'\xi'$$

In our approximation the arcs $HK$ and $H'K'$ may be considered straight lines. Hence we may write

$$\varphi = \frac{KK'}{HK'} = \frac{k\lambda}{D} \qquad \varphi' = \frac{HH'}{HK'} = \frac{k\lambda}{D}$$

where $D$ is the diameter of the lens. Consequently

$$\varphi = \varphi' \qquad \text{and} \qquad \frac{y}{x} = \frac{y'}{x'}$$

If we take account of the signs, we may write

$$-\frac{y'}{y} = \frac{x'}{x}$$

This is the equation for the *lateral magnification* of a lens immersed in a homogeneous medium. It too is identical with the equation for curved mirrors (§ 312).

**322.** Our equations for mirrors and lenses were derived under rather special conditions, like those illustrated in the Figures that aided our computations in the preceding Sections. Yet actually these equations are valid in the most diverse cases, both those of *converging* lenses (which are thicker at the center than at the edge) and *diverging* lenses (which are thicker at the edge than at the center) as well as concave mirrors and convex mirrors.

The individual examination of all these cases offers no difficulty, since the basic reasoning has already been set forth. The conclusion is always the same, whether it concerns the relation between the vergences or the lateral magnification. Differences are encountered only in the sign to be given to the various magnitudes that appear in these equations, such as $x$, $x'$, $f$, $r$, $y$, $y'$ (and possibly $\xi$, $\xi'$, $\varphi$).

The equations have general validity when $f$ (and $\varphi$) are taken as positive for concave mirrors and converging lenses (which are also called *positive*) and as negative for convex mirrors and diverging lenses (which are also called *negative*). Starting from the mirror or lens and proceeding toward the source, the measurement of $x$ (and $\xi$) is positive in the direction opposite to that in which the waves are propagated. Starting again from the mirror or lens, the measurement of $x'$ (and $\xi'$) is positive in the same direction as that in which the waves are propagated. The measurement of $y$ and $y'$ is positive on one side of the axis, and negative on the opposite side. The rule given for $f$ holds also for $r$, the radius of curvature of mirrors. For the radii of the surfaces of lenses, suitable conventions are adopted to make $f$ conform to the same rule.

In this discussion many terms occur that have a distinctively optical meaning. Examples are mirror, lens, and prism; we may add $f$ = focal length, $\varphi$ = power and also magnification, reflection, refraction, dispersion, etc. These expressions came into use for the first time in the study of the visual waves, and only for this reason do they have a traditionally optical character. But there is nothing exclusively optical about them. They refer to properties and phenomena characteristic of all wave propagations. Thus, people talk about acoustical mirrors, acoustical lenses, acoustical reflection, refraction and dispersion, etc.

These properties and phenomena have been examined in this Appendix, I should again like to make clear, as a study of what may be called general wave motion, which is applicable also to optics. But to make this application, the eye must be brought into play. That is exactly what I tried to do in the Chapters concerning the "science of vision."

**323.** What was said in § 322 may with greater force be repeated for the group of phenomena known as *diffraction*. Many people regard these phenomena as typically optical, whereas they are in fact common to wave motions of any sort. They occur whenever a wave front is limited by a diaphragm or interrupted by an obstruction.

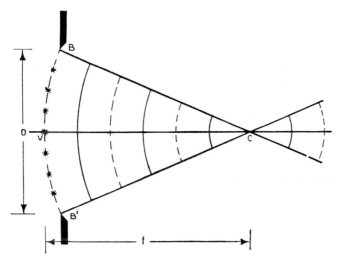

Fig. 124. Convergent spherical wave passing through a diaphragm

As phenomena, they are very complex. To study them with utmost rigor requires a rather intricate mathematical treatment. This is justifiable in a work intended to verify the correspondence between theory and experiment. But it would be inappropriate in a place such as this, whose purpose is only to offer a rapid and simple demonstration of the wave phenomena used in the earlier Chapters.

Accordingly, of the many diffraction phenomena known today, I shall confine myself here to analyzing only one. Moreover, I shall employ a very elementary method. The results to which it leads disagree with those obtained by more accurate investigations only in some minor differences in the numerical coefficients. The reasoning that I shall set forth allows the development of the phenomenon to be followed readily, so that the reader can understand its true nature.

The quickest way to reach the conclusions of interest to us is to analyze the behavior of waves on a surface, when an obstruction with an aperture of width $D$ is placed in their path. To take up at once the case to which I said I planned to restrict this discussion, I assume that the wave is convergent and that its center $C$ is situated on the axis of the aperture $BB'$ (Fig. 124). Hence the wave front reaches the points $B$ and $B'$ at the same instant. The point $V$, at the center of the arc $BB'$, is the vertex of the wave. Its radius of curvature is the distance $VC = f$.

**324.** What happens behind a diaphragm that limits the aperture of a wave front is studied by developing a concept explained in

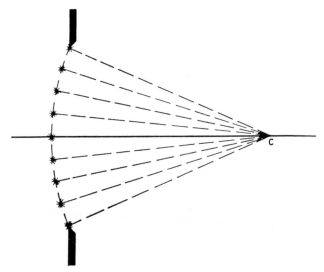

Fig. 125. Calculating the intensity of vibration at the center of a convergent spherical wave

§ 306. It was there summarized in the rule that the motion in a point outside a wave is the integral of all the motions generated individually in the point by all the secondary wavelets emitted by the points of the wave front.

Begin with the point $C$ (Fig. 125). By definition, the points of a wave front vibrate in phase. Hence the individual vibrations sent by each of them to $C$ must all be in phase, because $C$ is equidistant from them all. These vibrations, therefore, give rise to a vibration whose amplitude is the arithmetical sum of the amplitudes of the secondary vibrations (§ 298). This sum is an absolute maximum.

Now consider another point $P$, situated on $PC$, which is perpendicular to the axis of the diaphragm (Fig. 126a.) $P$ is chosen so that the distance $PB$ is less than $PB'$ by exactly one $\lambda$. We must sum all the motions that are produced at the point $P$ by the individual secondary wavelets. This sum is easily shown to be zero.

Draw the arc $BB''$ with its center at $P$. The segment $B'B''$ is exactly equal to $\lambda$, and the segment $VV'$ equals $\lambda/2$. Since $VP$ differs from $BP$ by $\lambda/2$, $B$ and $V$ as centers of secondary wavelets send vibrations in opposite phase to $P$. If the amplitude is constant on the wave front, the sum of these two secondary vibrations is zero (§ 298).

Moreover, for every elementary source on the arc $BV$ there is a corresponding source on the arc $VB'$ such that the two vibrations sent by them to $P$ are in opposite phase, and therefore yield a sum of zero. Consequently the total sum is zero. As a result, the point $P$ remains at rest.

**325.** The conclusions reached thus far (maximum movement in $C$ and no movement in $P$) are quite obvious. Less obvious are the conclusions reached when we consider other points on $PC$.

For instance, take a point $P$ at such a distance from $C$ that $PB' - PB = \lambda/2$ (Fig. 126b). The vibration that reaches $P$ from $B'$ is in opposite phase to the vibration arriving from $B$, and their sum is zero. But the sum of the other vibrations at $P$ is certainly not zero, because there are no other pairs of vibrations in opposite phase.

Now move the point $P$ a little farther away from $C$ (Fig. 126c). The secondary vibration arriving from $A_1$ is in opposition to that coming from $B$, and all the vibrations from points on the front $BA_2$ encounter others in opposition that come from the points on the front $A_1B'$, so that the sum of all these is zero. But the vibrations coming from the front $A_1A_2$ have no counterpart and yield a sum other than zero. This sum is smaller, however, than that accumu-

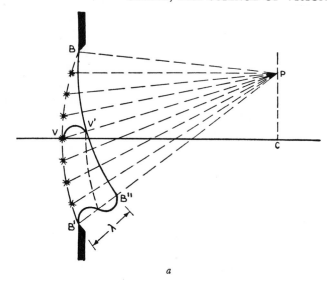

*a*

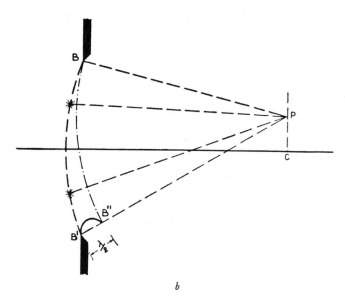

*b*

Fig. 126. Calculating the intensity of vibration at a point

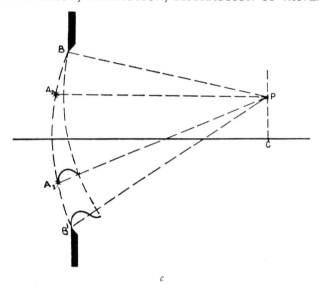

*c*

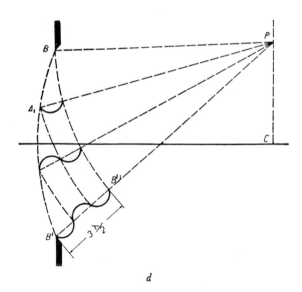

*d*

not on the axis of a segment of a convergent spherical wave

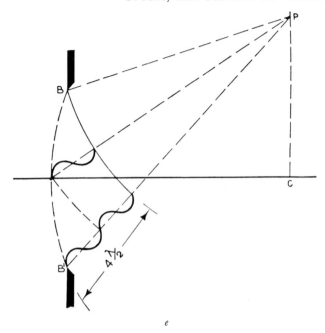

$e$

Fig. 126. (*Continued*)

lated at the position of $P$ in Fig. 126$b$, and decreases as $P$ recedes from $C$ to approach the position shown in Fig. 126$a$.

We conclude, then, with regard to points on the perpendicular to the axis at $C$, that as they withdraw from $C$, their motion steadily drops from a maximum at $C$ to zero at the position illustrated in Fig. 126$a$. This diminution of course occurs whether $P$ is on one side of the axis or the other.

Continue to move the point $P$ farther away from the axis, until it reaches the position defined by the difference

$$PB' - PB = 3\,\lambda/2$$

Fig. 126$d$ shows at a glance how to proceed in this case. With center at $P$, draw three arcs: $BB''$ with radius $PB$, a second arc with radius $PB + \lambda/2$, and the third with radius $PB + \lambda$. A comparison with Fig. 126$a$ shows that the sum of the effects produced at $P$ by the elementary sources distributed over the arc $A_1B'$ is zero. But the sum of the effects produced at $P$ by the elementary sources of the front $BA_1$ is other than zero. Therefore $P$ is again in motion. But its motion is no longer like that which it had when it was in positions close to $C$, because now the effective front is reduced to a third.

Fig. 126$e$ is very explicit. When $P$ is so far away from $C$ that

$$PB' - PB = 4\,\lambda/2$$

the sum of the actions at $P$ is zero. And so the cycle repeats itself.

**326.** We have seen that there is an absolute maximum of motion at the point $C$ on the axis. On the perpendicular to the axis at $C$, at points such that

$$PB' - PB = 2k\,\lambda/2 \qquad (k = \pm 1, 2, 3, \ldots)$$

the resultant motion is zero. But at intermediate points there is a

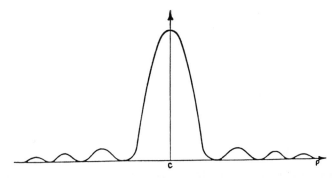

Fig. 127. Distribution curve of the intensity of vibration around the center of a convergent spherical wave

vibratory motion whose amplitude is other than zero and attains secondary maxima at points defined by

$$PB' - PB = (2k + 1)\,\lambda/2 \qquad (k = \pm 1, 2, 3, \ldots)$$

This phenomenon may be illustrated graphically by assigning the distances $PC$ to the abscissas, and to the ordinates the values of the amplitude of the motion at the corresponding point $P$ (Fig. 127). The curve thus obtained represents the distribution of the motion around $C$, the wave's center of curvature, and may be called the *curve of the centric.*

**327.** In this connection we may write several equations that are extensively applied. Fig. 128 generalizes the conditions discussed in §§ 324–326. The distance of $P$ from $C$ is denoted by $r$; the distance $B'B''$ is equated with $k\lambda$; the angle formed by $PV'$ with the axis is indicated by $\gamma$, which is taken to be equal to the angle formed by the two arcs $BB'$ and $BB''$, considered as straight lines; the

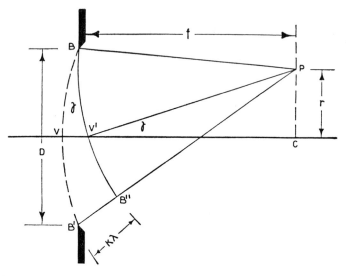

FIG. 128. Calculating the fundamental elements of the distribution curve shown in Fig. 127

distance from $C$ to the diaphragm is designated by $f$; and $D$ stands for the width of the aperture. The following relations are obvious:

$$\gamma = \frac{r}{f} = \frac{k\lambda}{D}$$

Hence

$$r = k\,\frac{\lambda}{D}\,f$$

Of all the values of $r$, the most interesting is that connected with the first minimum, as illustrated in Fig. 126a. The following equations correspond to this value

$$\gamma = \frac{\lambda}{D} \qquad r = \frac{\lambda}{D}f$$

**328.** The extension of the foregoing results to waves in space would not be simple if we wished to proceed with the greatest possible rigor. We should have to consider a spherical wave front, limited by a diaphragm with an aperture. Then we should have to go on to integrate the secondary wavelets at any point beyond the diaphragm. This can be done by means of integral calculus, when the hole in the diaphragm has a regular edge (circular, square, rectangular, or the like).

But such precise and intricate work is, in my opinion, not justifiable in this book. I shall instead confine myself to carrying over the conclusions utilized in previous Chapters.

When the hole in the diaphragm has a circular edge, and the center of the spherical wave lies on the axis of the hole, the results already obtained may be repeated for every section made by passing a plane through the axis and through a diameter of the hole. In other words, the constructions illustrated in Figs. 125–128 can be rotated around $CV$, the axis of the diaphragm.

Then in the plane passing through $C$ at right angles to the axis, the motion of the points presents an absolute maximum of amplitude in $C$, and drops steadily down to a circle, in whose points the amplitude is zero. The radius $r$ of this circle is given by

$$r = \frac{1.22\lambda}{D} f$$

and at $V$, the center of the diaphragm, it subtends the angle

$$\gamma = \frac{1.22\lambda}{D}$$

This circle encloses what is called the *central disk of the centric*. Outside the circle we find a ring of motion, and then another circumference of points at rest; then another ring of points in motion, and another circumference of points at rest, and so forth. The amplitude of the motion as distributed along a diameter is shown by the curve in Fig. 127.

# *Index*

◇◇◇◇◇◇◇◇◇◇◇◇

A CATALOG OF SELECTED
# DOVER BOOKS
## IN SCIENCE AND MATHEMATICS

# DOVER BOOKS

## IN SCIENCE AND MATHEMATICS

QUALITATIVE THEORY OF DIFFERENTIAL EQUATIONS, V.V. Nemytskii and V.V. Stepanov. Classic graduate-level text by two prominent Soviet mathematicians covers classical differential equations as well as topological dynamics and erqodic theory. Bibliographies. 523pp. 5⅜ × 8½. 65954-2 Pa. $10.95

MATRICES AND LINEAR ALGEBRA, Hans Schneider and George Phillip Barker. Basic textbook covers theory of matrices and its applications to systems of linear equations and related topics such as determinants, eigenvalues and differential equations. Numerous exercises. 432pp. 5⅜ × 8½. 66014-1 Pa. $8.95

QUANTUM THEORY, David Bohm. This advanced undergraduate-level text presents the quantum theory in terms of qualitative and imaginative concepts, followed by specific applications worked out in mathematical detail. Preface. Index. 655pp. 5⅜ × 8½. 65969-0 Pa. $10.95

ATOMIC PHYSICS (8th edition), Max Born. Nobel laureate's lucid treatment of kinetic theory of gases, elementary particles, nuclear atom, wave-corpuscles, atomic structure and spectral lines, much more. Over 40 appendices, bibliography. 495pp. 5⅜ × 8½. 65984-4 Pa. $11.95

ELECTRONIC STRUCTURE AND THE PROPERTIES OF SOLIDS: The Physics of the Chemical Bond, Walter A. Harrison. Innovative text offers basic understanding of the electronic structure of covalent and ionic solids, simple metals, transition metals and their compounds. Problems. 1980 edition. 582pp. 6⅛ × 9¼. 66021-4 Pa. $14.95

BOUNDARY VALUE PROBLEMS OF HEAT CONDUCTION, M. Necati Özisik. Systematic, comprehensive treatment of modern mathematical methods of solving problems in heat conduction and diffusion. Numerous examples and problems. Selected references. Appendices. 505pp. 5⅜ × 8½. 65990-9 Pa. $11.95

A SHORT HISTORY OF CHEMISTRY (3rd edition), J.R. Partington. Classic exposition explores origins of chemistry, alchemy, early medical chemistry, nature of atmosphere, theory of valency, laws and structure of atomic theory, much more. 428pp. 5⅜ × 8½. (Available in U.S. only) 65977-1 Pa. $10.95

A HISTORY OF ASTRONOMY, A. Pannekoek. Well-balanced, carefully reasoned study covers such topics as Ptolemaic theory, work of Copernicus, Kepler, Newton, Eddington's work on stars, much more. Illustrated. References. 521pp. 5⅜ × 8½. 65994-1 Pa. $11.95

PRINCIPLES OF METEOROLOGICAL ANALYSIS, Walter J. Saucier. Highly respected, abundantly illustrated classic reviews atmospheric variables, hydrostatics, static stability, various analyses (scalar, cross-section, isobaric, isentropic, more). For intermediate meteorology students. 454pp. 6⅛ × 9¼. 65979-8 Pa. $12.95

RELATIVITY, THERMODYNAMICS AND COSMOLOGY, Richard C. Tolman. Landmark study extends thermodynamics to special, general relativity; also applications of relativistic mechanics, thermodynamics to cosmological models. 501pp. 5⅜ × 8½. 65383-8 Pa. $11.95

APPLIED ANALYSIS, Cornelius Lanczos. Classic work on analysis and design of finite processes for approximating solution of analytical problems. Algebraic equations, matrices, harmonic analysis, quadrature methods, much more. 559pp. 5⅜ × 8½. 65656-X Pa. $11.95

SPECIAL RELATIVITY FOR PHYSICISTS, G. Stephenson and C.W. Kilmister. Concise elegant account for nonspecialists. Lorentz transformation, optical and dynamical applications, more. Bibliography. 108pp. 5⅜ × 8½. 65519-9 Pa. $3.95

INTRODUCTION TO ANALYSIS, Maxwell Rosenlicht. Unusually clear, accessible coverage of set theory, real number system, metric spaces, continuous functions, Riemann integration, multiple integrals, more. Wide range of problems. Undergraduate level. Bibliography. 254pp. 5⅜ × 8½. 65038-3 Pa. $7.00

INTRODUCTION TO QUANTUM MECHANICS With Applications to Chemistry, Linus Pauling & E. Bright Wilson, Jr. Classic undergraduate text by Nobel Prize winner applies quantum mechanics to chemical and physical problems. Numerous tables and figures enhance the text. Chapter bibliographies. Appendices. Index. 468pp. 5⅜ × 8½. 64871-0 Pa. $9.95

ASYMPTOTIC EXPANSIONS OF INTEGRALS, Norman Bleistein & Richard A. Handelsman. Best introduction to important field with applications in a variety of scientific disciplines. New preface. Problems. Diagrams. Tables. Bibliography. Index. 448pp. 5⅜ × 8½. 65082-0 Pa. $10.95

MATHEMATICS APPLIED TO CONTINUUM MECHANICS, Lee A. Segel. Analyzes models of fluid flow and solid deformation. For upper-level math, science and engineering students. 608pp. 5⅜ × 8½. 65369-2 Pa. $12.95

ELEMENTS OF REAL ANALYSIS, David A. Sprecher. Classic text covers fundamental concepts, real number system, point sets, functions of a real variable, Fourier series, much more. Over 500 exercises. 352pp. 5⅜ × 8½. 65385-4 Pa. $8.95

PHYSICAL PRINCIPLES OF THE QUANTUM THEORY, Werner Heisenberg. Nobel Laureate discusses quantum theory, uncertainty, wave mechanics, work of Dirac, Schroedinger, Compton, Wilson, Einstein, etc. 184pp. 5⅜ × 8½. 60113-7 Pa. $4.95

INTRODUCTORY REAL ANALYSIS, A.N. Kolmogorov, S.V. Fomin. Translated by Richard A. Silverman. Self-contained, evenly paced introduction to real and functional analysis. Some 350 problems. 403pp. 5⅜ × 8½. 61226-0 Pa. $7.95

PROBLEMS AND SOLUTIONS IN QUANTUM CHEMISTRY AND PHYSICS, Charles S. Johnson, Jr. and Lee G. Pedersen. Unusually varied problems, detailed solutions in coverage of quantum mechanics, wave mechanics, angular momentum, molecular spectroscopy, scattering theory, more. 280 problems plus 139 supplementary exercises. 430pp. 6½ × 9¼. 65236-X Pa. $10.95

ASYMPTOTIC METHODS IN ANALYSIS, N.G. de Bruijn. An inexpensive, comprehensive guide to asymptotic methods—the pioneering work that teaches by explaining worked examples in detail. Index. 224pp. 5⅜ × 8½.    64221-6 Pa. $5.95

OPTICAL RESONANCE AND TWO-LEVEL ATOMS, L. Allen and J.H. Eberly. Clear, comprehensive introduction to basic principles behind all quantum optical resonance phenomena. 53 illustrations. Preface. Index. 256pp. 5⅜ × 8½.
65533-4 Pa. $6.95

COMPLEX VARIABLES, Francis J. Flanigan. Unusual approach, delaying complex algebra till harmonic functions have been analyzed from real variable viewpoint. Includes problems with answers. 364pp. 5⅜ × 8½.    61388-7 Pa. $7.95

ATOMIC SPECTRA AND ATOMIC STRUCTURE, Gerhard Herzberg. One of best introductions; especially for specialist in other fields. Treatment is physical rather than mathematical. 80 illustrations. 257pp. 5⅜ × 8½.    60115-3 Pa. $4.95

APPLIED COMPLEX VARIABLES, John W. Dettman. Step-by-step coverage of fundamentals of analytic function theory—plus lucid exposition of 5 important applications: Potential Theory; Ordinary Differential Equations; Fourier Transforms; Laplace Transforms; Asymptotic Expansions. 66 figures. Exercises at chapter ends. 512pp. 5⅜ × 8½.    64670-X Pa. $10.95

ULTRASONIC ABSORPTION: An Introduction to the Theory of Sound Absorption and Dispersion in Gases, Liquids and Solids, A.B. Bhatia. Standard reference in the field provides a clear, systematically organized introductory review of fundamental concepts for advanced graduate students, research workers. Numerous diagrams. Bibliography. 440pp. 5⅜ × 8½.    64917-2 Pa. $8.95

UNBOUNDED LINEAR OPERATORS: Theory and Applications, Seymour Goldberg. Classic presents systematic treatment of the theory of unbounded linear operators in normed linear spaces with applications to differential equations. Bibliography. 199pp. 5⅜ × 8½.    64830-3 Pa. $7.00

LIGHT SCATTERING BY SMALL PARTICLES, H.C. van de Hulst. Comprehensive treatment including full range of useful approximation methods for researchers in chemistry, meteorology and astronomy. 44 illustrations. 470pp. 5⅜ × 8½.    64228-3 Pa. $9.95

CONFORMAL MAPPING ON RIEMANN SURFACES, Harvey Cohn. Lucid, insightful book presents ideal coverage of subject. 334 exercises make book perfect for self-study. 55 figures. 352pp. 5⅜ × 8¼.    64025-6 Pa. $8.95

OPTICKS, Sir Isaac Newton. Newton's own experiments with spectroscopy, colors, lenses, reflection, refraction, etc., in language the layman can follow. Foreword by Albert Einstein. 532pp. 5⅜ × 8½.    60205-2 Pa. $8.95

GENERALIZED INTEGRAL TRANSFORMATIONS, A.H. Zemanian. Graduate-level study of recent generalizations of the Laplace, Mellin, Hankel, K. Weierstrass, convolution and other simple transformations. Bibliography. 320pp. 5⅜ × 8½.    65375-7 Pa. $7.95

THE ELECTROMAGNETIC FIELD, Albert Shadowitz. Comprehensive undergraduate text covers basics of electric and magnetic fields, builds up to electromagnetic theory. Also related topics, including relativity. Over 900 problems. 768pp. 5⅜ × 8¼. 65660-8 Pa. $15.95

FOURIER SERIES, Georgi P. Tolstov. Translated by Richard A. Silverman. A valuable addition to the literature on the subject, moving clearly from subject to subject and theorem to theorem. 107 problems, answers. 336pp. 5⅜ × 8½. 63317-9 Pa. $7.95

THEORY OF ELECTROMAGNETIC WAVE PROPAGATION, Charles Herach Papas. Graduate-level study discusses the Maxwell field equations, radiation from wire antennas, the Doppler effect and more. xiii + 244pp. 5⅜ × 8½. 65678-0 Pa. $6.95

DISTRIBUTION THEORY AND TRANSFORM ANALYSIS: An Introduction to Generalized Functions, with Applications, A.H. Zemanian. Provides basics of distribution theory, describes generalized Fourier and Laplace transformations. Numerous problems. 384pp. 5⅜ × 8½. 65479-6 Pa. $8.95

THE PHYSICS OF WAVES, William C. Elmore and Mark A. Heald. Unique overview of classical wave theory. Acoustics, optics, electromagnetic radiation, more. Ideal as classroom text or for self-study. Problems. 477pp. 5⅜ × 8½. 64926-1 Pa. $10.95

CALCULUS OF VARIATIONS WITH APPLICATIONS, George M. Ewing. Applications-oriented introduction to variational theory develops insight and promotes understanding of specialized books, research papers. Suitable for advanced undergraduate/graduate students as primary, supplementary text. 352pp. 5⅜ × 8½. 64856-7 Pa. $8.50

A TREATISE ON ELECTRICITY AND MAGNETISM, James Clerk Maxwell. Important foundation work of modern physics. Brings to final form Maxwell's theory of electromagnetism and rigorously derives his general equations of field theory. 1,084pp. 5⅜ × 8½. 60636-8, 60637-6 Pa., Two-vol. set $19.00

AN INTRODUCTION TO THE CALCULUS OF VARIATIONS, Charles Fox. Graduate-level text covers variations of an integral, isoperimetrical problems, least action, special relativity, approximations, more. References. 279pp. 5⅜ × 8½. 65499-0 Pa. $6.95

HYDRODYNAMIC AND HYDROMAGNETIC STABILITY, S. Chandrasekhar. Lucid examination of the Rayleigh-Benard problem; clear coverage of the theory of instabilities causing convection. 704pp. 5⅜ × 8¼. 64071-X Pa. $12.95

CALCULUS OF VARIATIONS, Robert Weinstock. Basic introduction covering isoperimetric problems, theory of elasticity, quantum mechanics, electrostatics, etc. Exercises throughout. 326pp. 5⅜ × 8½. 63069-2 Pa. $7.95

DYNAMICS OF FLUIDS IN POROUS MEDIA, Jacob Bear. For advanced students of ground water hydrology, soil mechanics and physics, drainage and irrigation engineering and more. 335 illustrations. Exercises, with answers. 784pp. 6⅛ × 9¼. 65675-6 Pa. $19.95

THE FOUR-COLOR PROBLEM: Assaults and Conquest, Thomas L. Saaty and Paul G. Kainen. Engrossing, comprehensive account of the century-old combinatorial topological problem, its history and solution. Bibliographies. Index. 110 figures. 228pp. 5⅜ × 8½. 65092-8 Pa. $6.00

CATALYSIS IN CHEMISTRY AND ENZYMOLOGY, William P. Jencks. Exceptionally clear coverage of mechanisms for catalysis, forces in aqueous solution, carbonyl- and acyl-group reactions, practical kinetics, more. 864pp. 5⅜ × 8½. 65460-5 Pa. $18.95

PROBABILITY: An Introduction, Samuel Goldberg. Excellent basic text covers set theory, probability theory for finite sample spaces, binomial theorem, much more. 360 problems. Bibliographies. 322pp. 5⅜ × 8½. 65252-1 Pa. $7.95

LIGHTNING, Martin A. Uman. Revised, updated edition of classic work on the physics of lightning. Phenomena, terminology, measurement, photography, spectroscopy, thunder, more. Reviews recent research. Bibliography. Indices. 320pp. 5⅜ × 8¼. 64575-4 Pa. $7.95

PROBABILITY THEORY: A Concise Course, Y.A. Rozanov. Highly readable, self-contained introduction covers combination of events, dependent events, Bernoulli trials, etc. Translation by Richard Silverman. 148pp. 5⅜ × 8¼.
63544-9 Pa. $4.50

THE CEASELESS WIND: An Introduction to the Theory of Atmospheric Motion, John A. Dutton. Acclaimed text integrates disciplines of mathematics and physics for full understanding of dynamics of atmospheric motion. Over 400 problems. Index. 97 illustrations. 640pp. 6 × 9. 65096-0 Pa. $16.95

STATISTICS MANUAL, Edwin L. Crow, et al. Comprehensive, practical collection of classical and modern methods prepared by U.S. Naval Ordnance Test Station. Stress on use. Basics of statistics assumed. 288pp. 5⅜ × 8½.
60599-X Pa. $6.00

WIND WAVES: Their Generation and Propagation on the Ocean Surface, Blair Kinsman. Classic of oceanography offers detailed discussion of stochastic processes and power spectral analysis that revolutionized ocean wave theory. Rigorous, lucid. 676pp. 5⅜ × 8½. 64652-1 Pa. $14.95

STATISTICAL METHOD FROM THE VIEWPOINT OF QUALITY CONTROL, Walter A. Shewhart. Important text explains regulation of variables, uses of statistical control to achieve quality control in industry, agriculture, other areas. 192pp. 5⅜ × 8½. 65232-7 Pa. $6.00

THE INTERPRETATION OF GEOLOGICAL PHASE DIAGRAMS, Ernest G. Ehlers. Clear, concise text emphasizes diagrams of systems under fluid or containing pressure; also coverage of complex binary systems, hydrothermal melting, more. 288pp. 6½ × 9¼. 65389-7 Pa. $8.95

STATISTICAL ADJUSTMENT OF DATA, W. Edwards Deming. Introduction to basic concepts of statistics, curve fitting, least squares solution, conditions without parameter, conditions containing parameters. 26 exercises worked out. 271pp. 5⅜ × 8½. 64685-8 Pa. $7.95

DE RE METALLICA, Georgius Agricola. The famous Hoover translation of greatest treatise on technological chemistry, engineering, geology, mining of early modern times (1556). All 289 original woodcuts. 638pp. 6¾ × 11.
60006-8 Clothbd. $15.95

SOME THEORY OF SAMPLING, William Edwards Deming. Analysis of the problems, theory and design of sampling techniques for social scientists, industrial managers and others who find statistics increasingly important in their work. 61 tables. 90 figures. xvii + 602pp. 5⅜ × 8½.
64684-X Pa. $14.95

THE VARIOUS AND INGENIOUS MACHINES OF AGOSTINO RAMELLI: A Classic Sixteenth-Century Illustrated Treatise on Technology, Agostino Ramelli. One of the most widely known and copied works on machinery in the 16th century. 194 detailed plates of water pumps, grain mills, cranes, more. 608pp. 9 × 12.
25497-6 Clothbd. $34.95

LINEAR PROGRAMMING AND ECONOMIC ANALYSIS, Robert Dorfman, Paul A. Samuelson and Robert M. Solow. First comprehensive treatment of linear programming in standard economic analysis. Game theory, modern welfare economics, Leontief input-output, more. 525pp. 5⅜ × 8½.
65491-5 Pa. $12.95

ELEMENTARY DECISION THEORY, Herman Chernoff and Lincoln E. Moses. Clear introduction to statistics and statistical theory covers data processing, probability and random variables, testing hypotheses, much more. Exercises. 364pp. 5⅜ × 8½.
65218-1 Pa. $8.95

THE COMPLEAT STRATEGYST: Being a Primer on the Theory of Games of Strategy, J.D. Williams. Highly entertaining classic describes, with many illustrated examples, how to select best strategies in conflict situations. Prefaces. Appendices. 268pp. 5⅜ × 8½.
25101-2 Pa. $5.95

MATHEMATICAL METHODS OF OPERATIONS RESEARCH, Thomas L. Saaty. Classic graduate-level text covers historical background, classical methods of forming models, optimization, game theory, probability, queueing theory, much more. Exercises. Bibliography. 448pp. 5⅜ × 8¼.
65703-5 Pa. $12.95

CONSTRUCTIONS AND COMBINATORIAL PROBLEMS IN DESIGN OF EXPERIMENTS, Damaraju Raghavarao. In-depth reference work examines orthogonal Latin squares, incomplete block designs, tactical configuration, partial geometry, much more. Abundant explanations, examples. 416pp. 5⅜ × 8¼.
65685-3 Pa. $10.95

THE ABSOLUTE DIFFERENTIAL CALCULUS (CALCULUS OF TENSORS), Tullio Levi-Civita. Great 20th-century mathematician's classic work on material necessary for mathematical grasp of theory of relativity. 452pp. 5⅜ × 8½.
63401-9 Pa. $9.95

VECTOR AND TENSOR ANALYSIS WITH APPLICATIONS, A.I. Borisenko and I.E. Tarapov. Concise introduction. Worked-out problems, solutions, exercises. 257pp. 5⅜ × 8¼.
63833-2 Pa. $6.95

TENSOR CALCULUS, J.L. Synge and A. Schild. Widely used introductory text covers spaces and tensors, basic operations in Riemannian space, non-Riemannian spaces, etc. 324pp. 5⅜ × 8¼. 63612-7 Pa. $7.00

A CONCISE HISTORY OF MATHEMATICS, Dirk J. Struik. The best brief history of mathematics. Stresses origins and covers every major figure from ancient Near East to 19th century. 41 illustrations. 195pp. 5⅜ × 8½. 60255-9 Pa. $7.95

A SHORT ACCOUNT OF THE HISTORY OF MATHEMATICS, W.W. Rouse Ball. One of clearest, most authoritative surveys from the Egyptians and Phoenicians through 19th-century figures such as Grassman, Galois, Riemann. Fourth edition. 522pp. 5⅜ × 8½. 20630-0 Pa. $9.95

HISTORY OF MATHEMATICS, David E. Smith. Non-technical survey from ancient Greece and Orient to late 19th century; evolution of arithmetic, geometry, trigonometry, calculating devices, algebra, the calculus. 362 illustrations. 1,355pp. 5⅜ × 8½. 20429-4, 20430-8 Pa., Two-vol. set $21.90

THE GEOMETRY OF RENÉ DESCARTES, René Descartes. The great work founded analytical geometry. Original French text, Descartes' own diagrams, together with definitive Smith-Latham translation. 244pp. 5⅜ × 8½.
60068-8 Pa. $6.00

THE ORIGINS OF THE INFINITESIMAL CALCULUS, Margaret E. Baron. Only fully detailed and documented account of crucial discipline: origins; development by Galileo, Kepler, Cavalieri; contributions of Newton, Leibniz, more. 304pp. 5⅜ × 8½. (Available in U.S. and Canada only) 65371-4 Pa. $7.95

THE HISTORY OF THE CALCULUS AND ITS CONCEPTUAL DEVELOP-MENT, Carl B. Boyer. Origins in antiquity, medieval contributions, work of Newton, Leibniz, rigorous formulation. Treatment is verbal. 346pp. 5⅜ × 8½.
60509-4 Pa. $6.95

THE THIRTEEN BOOKS OF EUCLID'S ELEMENTS, translated with introduction and commentary by Sir Thomas L. Heath. Definitive edition. Textual and linguistic notes, mathematical analysis. 2500 years of critical commentary. Not abridged. 1,414pp. 5⅜ × 8½. 60088-2, 60089-0, 60090-4 Pa., Three-vol. set $26.85

A HISTORY OF VECTOR ANALYSIS: The Evolution of the Idea of a Vectorial System, Michael J. Crowe. The first large-scale study of the history of vector analysis, now the standard on the subject. Unabridged republication of the edition published by University of Notre Dame Press, 1967, with second preface by Michael C. Crowe. Index. 278pp. 5⅜ × 8½. 64955-5 Pa. $7.00

THE HISTORICAL ROOTS OF ELEMENTARY MATHEMATICS, Lucas N.H. Bunt, Phillip S. Jones, and Jack D. Bedient. Fundamental underpinnings of modern arithmetic, algebra, geometry and number systems derived from ancient civilizations. 320pp. 5⅜ × 8½. 25563-8 Pa. $7.95

CALCULUS REFRESHER FOR TECHNICAL PEOPLE, A. Albert Klaf. Covers important aspects of integral and differential calculus via 756 questions. 566 problems, most answered. 431pp. 5⅜ × 8½. 20370-0 Pa. $7.95

CHALLENGING MATHEMATICAL PROBLEMS WITH ELEMENTARY SOLUTIONS, A.M. Yaglom and I.M. Yaglom. Over 170 challenging problems on probability theory, combinatorial analysis, points and lines, topology, convex polygons, many other topics. Solutions. Total of 445pp. 5⅜ × 8½. Two-vol. set.

Vol. I 65536-9 Pa. $5.95
Vol. II 65537-7 Pa. $5.95

FIFTY CHALLENGING PROBLEMS IN PROBABILITY WITH SOLUTIONS, Frederick Mosteller. Remarkable puzzlers, graded in difficulty, illustrate elementary and advanced aspects of probability. Detailed solutions. 88pp. 5⅜ × 8½.
65355-2 Pa. $3.95

EXPERIMENTS IN TOPOLOGY, Stephen Barr. Classic, lively explanation of one of the byways of mathematics. Klein bottles, Moebius strips, projective planes, map coloring, problem of the Koenigsberg bridges, much more, described with clarity and wit. 43 figures. 210pp. 5⅜ × 8½. 25933-1 Pa. $4.95

RELATIVITY IN ILLUSTRATIONS, Jacob T. Schwartz. Clear non-technical treatment makes relativity more accessible than ever before. Over 60 drawings illustrate concepts more clearly than text alone. Only high school geometry needed. Bibliography. 128pp. 6⅛ × 9¼. 25965-X Pa. $5.95

AN INTRODUCTION TO ORDINARY DIFFERENTIAL EQUATIONS, Earl A. Coddington. A thorough and systematic first course in elementary differential equations for undergraduates in mathematics and science, with many exercises and problems (with answers). Index. 304pp. 5⅜ × 8¼. 65942-9 Pa. $7.95

FOURIER SERIES AND ORTHOGONAL FUNCTIONS, Harry F. Davis. An incisive text combining theory and practical example to introduce Fourier series, orthogonal functions and applications of the Fourier method to boundary-value problems. 570 exercises. Answers and notes. 416pp. 5⅜ × 8½. 65973-9 Pa. $8.95

THE THOERY OF BRANCHING PROCESSES, Theodore E. Harris. First systematic, comprehensive treatment of branching (i.e. multiplicative) processes and their applications. Galton-Watson model, Markov branching processes, electron-photon cascade, many other topics. Rigorous proofs. Bibliography. 240pp. 5⅜ × 8½. 65952-6 Pa. $6.95

AN INTRODUCTION TO ALGEBRAIC STRUCTURES, Joseph Landin. Superb self-contained text covers "abstract algebra": sets and numbers, theory of groups, theory of rings, much more. Numerous well-chosen examples, exercises. 247pp. 5⅜ × 8½. 65940-2 Pa. $6.95

GAMES AND DECISIONS: Introduction and Critical Survey, R. Duncan Luce and Howard Raiffa. Superb non-technical introduction to game theory, primarily applied to social sciences. Utility theory, zero-sum games, n-person games, decision-making, much more. Bibliography. 509pp. 5⅜ × 8½. 65943-7 Pa. $10.95

---